MATHEMATICAL BAFFLERS

Edited by
ANGELA DUNN

Illustrations by Ed Kysar

Dover Publications, Inc.
New York

This Dover edition, first published in 1980, is a revised and cor-
rected republication of the work originally published by the
McGraw-Hill Book Company in 1964. A new Preface, replacing
the original Foreword and Preface, has been prepared specially for
this edition by the editor.

International Standard Book Number

ISBN-13: 978-0-486-23961-3
ISBN-10: 0-486-23961-6

Library of Congress Catalog Card Number: 79-56810

Manufactured in the United States by RR Donnelley
23961613   2015
www.doverpublications.com

PREFACE TO THE DOVER EDITION

This book is an outgrowth of one of the most successful corporate advertising campaigns in the history of technical publications, a weekly series called "Problematical Recreations," which ran for twelve years in *Aviation Week* magazine and the *Electronic News*, winning the top readership award year after year.

Conceived and sponsored by Litton Industries of Beverly Hills, California, "Problematical Recreations" offered a weekly mathematical puzzle, the answer appearing the following week, geared to attract the technically minded. Happily, the series proved to be more than well read; it was actually aided and perpetuated by the readers themselves.

The quality of their written response was the key to the series' continuing appeal. Week after week letters from engineers, mathematicians, scientists, and puzzle fans in general would offer a more elegant solution, or an interesting mathematical sidelight to a problem from our series. Often readers would challenge us for an explanation, and occasionally they would disagree, sometimes vehemently, with our published solution. But always they exhibited original thinking. It was the quantity of imaginative puzzle contributions—novel offerings poured in from all over the United States and from a dozen foreign countries—that kept the campaign going at a high level of interest for twelve years.

As director of "Problematical Recreations" from 1962 until its cancellation in 1971, I was fortunate in acquiring a staff of some of the best creative minds in mathematics to help check and evaluate each original contribution. My chief consultant, the late David L. Silverman of the University of California at Los Angeles, was truly a mathematical genius. His inexhaustible knowledge, his infinite supply of ingenious original puzzles, and his ability to communicate any principle or idea simply are responsible for both the series' success and this volume. One of David Silverman's many admirers, Mr. George Koch, president of Guidance Industries Corporation of San Francisco, commented: "He was the only mathematician I found in front of whom I was comfortable admitting ignorance. He answered my ignorance with information, not disdain, and thereby taught me a great deal."

I relied heavily on Mr. Silverman's expertise in handling the volume of correspondence. Each letter was answered personally, after careful checking and research, a fact which so surprised and pleased one reader in Washington, D.C., that he wrote me: "Thank you for not sending me the 'bed bug' letter. You present Litton as a warm and human organization."* Because this correspondence between myself and the readers of "Problematical Recreations" may enhance your enjoyment of a puzzle, shed new mathematical light, or simply amuse, selections have been included at the beginning of each of the seven sections of this book.

When the puzzles were originally published, their sequence was chosen to provide interesting variety from week to week. You will find, therefore, that the selections here run the gamut from simple problems requiring no mathematical background to those that would challenge a professional mathematician. For example, a little imagination is all that is required to solve the following sequence problem from Chapter 6:

What letter follows OTTFFSSE__?

---

*Bed bug letter: a form letter, from a company to an individual who has made a complaint, which promises to correct a situation, but is actually only intended to pacify the person making the objection.

On the other end of the scale, advanced mathematics is involved in solving a variation of "The Alpenstock" (first problem of Chapter 5), and an acquaintance with Number Theory is required for the problems in Chapter 7.

In making this selection of more than 150 posers, we chose those that we hope combine the unusual, the unexpected, and the nonobvious. You will find, therefore, that a majority of the solutions may be reached by the application of a well-conceived hunch rather than by drudgery and exhaustive checking of tables. For our object is, after all, to entertain.

The mathematical challenges that follow have been contributed by dozens of puzzlers throughout this country, and from all over the world, most of them skilled mathematicians and applied scientists. We share their pet brain twisters and original work with you in these pages. For consistently submitting original and ingenious puzzles, the editor is indebted to: Mr. Leonard A. Baljay of Cherry Hill, New Jersey; Mr. Walter Penney of Greenbelt, Maryland; Mr. Charles Baker of Los Angeles, California; Mr. Noel A. Longmore of Kent, England; Mr. B. van Blaricum of Melbourne, Australia; Mr. William Shooman of Orange, California; and Mr. J. N. A. Hawkins of Pacific Palisades, California.

Acknowledgement is made also to *American Mathematical Monthly* and *Mathematics Magazine,* the two publications of the Mathematical Association of America, for permission to include in the "Problematical Recreations" series a few sample problems from their large and excellent stock.

For their patient counseling and technical assistance in conducting the series, the editor is grateful to Dr. Silverman and Dr. Harry Lass of the California Institute of Technology.

This book is for those who take pleasure in the process of reasoning, who enjoy exercising their inventive faculties, who delight in the pursuit of an elusive proof. If the reader enjoys these particular challenges, he or she is indebted to all the gentlemen named above and to all those hardy fellows who took the time to write to "Problematical Recreations."

ANGELA DUNN

CONTENTS

1. SAY IT WITH LETTERS 1
Algebraic Amusements

2. AXIOMS, ANGLES, AND ARCS 49
Geometric Exercises

3. SOLVING IN INTEGERS 71
Diophantine Diversions

4. THE DATA SEEKERS 91
Problems in Logic and Deduction

5. MINDING YOUR P's AND Q's 135
Probability Posers

6. NOW YOU SEE IT 153
Insight Puzzles

7. PERMUTATIONS, PARTITIONS, AND PRIMES 181
Assorted Number Theory Problems

1

SAY IT WITH LETTERS

abcdefghijklmnopqrstuvwxyz abcdefghijklmnopqrstuvwxyz abcdefghijklmnopqrstuv

Algebraic Amusements

Four Fours (SOLUTION, Page 4)

An Inequality (SOLUTION, Page 10)

Variations on the Tricolor · *A Transposed Equation* · *A Ladder Problem* · *The Beauty Contest* · *The Two Motorists* · *Improving on an Escalator* · *An Age Problem* · *The Freight Train* · *A Close Race* · *The Watch Clocker* · *The Evasive Engineer* · *A Lifetime Task* · *The Ages of Man* · *An Infinite Product* · *The Tarry-Town Riddle* · *Three Generations* · *Renewing Old Ties* · *Another Age Problem* · *A Difficult Factorization* · *The Competing Airlines* · *The Optimal Salary* · *A Trigonometric Root* · *The Bricklayers* · *A Stockmarket Problem* · *The Family Legacy* · *A Matter of Great Interest* · *The Senior Prom* · *The Digital Atom Smasher* · *A Sure Thing* · *Two Men on a Horse* · *A Decreasing Ratio* · *An Infinitude of Twos*

SOLUTIONS: Page 41

FOUR FOURS

Using mathematical symbols to modify four fours it is possible to write expressions for all the numbers from 0 to 100, as well as millions of others. Example: $2 = 4/4 + 4/4$. In this manner arrange four fours to equal these progressively more difficult numbers: 13, 19, 33, 85.

Solution

$13 = 44/4 + \sqrt{4}$; $19 = 4! - 4 - 4/4$; $33 = 4/.\dot{4} + (\sqrt{4} + \sqrt{4})!$;
$85 = 4 + (4/.\dot{4})^{\sqrt{4}}$

There are other equally valid solutions to this problem. We chose to print this one which made use of a notation we found had fallen out of common use, judging from the response we received. The letters ranged from polite requests ("We would appreciate your clarifying the origin and current usage of the symbols used in your last two answers.") to desperate notes ("I can hold out no longer. Please explain the cryptic notation .$\dot{4}$ used to force a solution to your puzzle. I have been unable to find anyone who can explain"). ". . . how do you get $4/.\dot{4} =$ 9??? wrote a gentleman at M.I.T.'s Lincoln Laboratory. ". . . you use a symbol which is unfamiliar to myself and my associates. I am referring to the symbol .$\dot{4}$, . ." said a senior scientist at AVCO. ". . . .$\dot{4}$ must equal $4/9$ in this problem but I have never seen this notation used before . . ." declared an engineer from the System Development Corporation in Phoenix. From a researcher in Roseville, Michigan: "Please explain the dot over the .$\dot{4}$. I've called three (3) universities in Detroit and drew a blank."

We assured our readers that the dot over the digit, far from being a sophisticated code, is a symbol indicating repeating decimals. Thus .$\dot{4}$ = .444444444 ad infinitum or $4/9$. Our reference to readers who requested "a standard reference which will explain the operations involved in this nomenclature" was page 3005, Arbitrary Signs and Symbols, Webster's New International Dictionary, Second Edition. The symbol can also be found in numerous college algebra and number theory texts.

We mentioned alternate solutions earlier in this discussion. Here they are:

From Cleveland, Ohio: "$13 = \dfrac{4 - .4}{.4} + 4$; $33 = \dfrac{4 - .4}{.4} +$ $4!$; $85 = \dfrac{4! + .4}{.4} + 4!$"

4

These nifty solutions neatly avoid our dot over the digit as well as the symbol $\sqrt{}$. In connection with our use of the latter symbol, the sender continued, "Does not $\sqrt{}$ have an implied 2; $\sqrt[2]{}$? You're cheating and it isn't necessary." We saw it wasn't necessary, but we could not agree to the cheating charge, since, after all, 4.4 implies a ten (the number base), i.e., $(4.4)_{10}$. In any event the problem did not mention implications.

From an aeronautical engineer in Culver City, California, came the following unusual approach to the problem:

Solutions to your problem can also be readily obtained via use of the gamma or generalized factorial function "Γ," wherein

$$\Gamma(4) = 3! = 6$$

i.e.: $13 = \Gamma(4) \times \sqrt{4} + \dfrac{4}{4}$

$$19 = \Gamma(4)/\sqrt{4} + 4 \times 4$$
$$33 = \Gamma(4) + \Gamma(4)/\sqrt{4} + 4!$$
$$85 = (\Gamma(4)/\sqrt{4})^4 + 4$$

I believe that the Greek letter gamma (Γ) is more universally recognized as a mathematical symbol than is your usage of .4. What branch of mathematics expounds your inference that .$\dot{4}$ is the equivalent of 4/9?? I have asked several of the more eminent senior staff scientists here if they recognized the significance of the term .4. None of them had ever seen or heard of this form of mathematical expression. Three of these contacted scientists are Ph.D. mathematicians.

The writer added the following postscript: "Another simple solution for 13 is: $13 = \dfrac{4!}{4} + \dfrac{4}{4}$."

For those who are interested in such things (a surprisingly large number) we offer the following list of representations for all the integers from one to a hundred, submitted by Robert L. Johnsen, Jr., of Los Angeles:

0   $\dfrac{4}{4} - \dfrac{4}{4}$ 2 $\dfrac{4}{4} + \dfrac{4}{4}$

1   $\dfrac{4 \cdot 4}{4 \cdot 4}$ 3 $4 - \sqrt{4} + \dfrac{4}{4}$

4 $\quad \sqrt{\dfrac{4}{.4}} + \dfrac{4}{4}$

5 $\quad \sqrt{\dfrac{4}{.4}} + \dfrac{4}{\sqrt{4}}$

6 $\quad \dfrac{4!4}{4 \cdot 4}$

7 $\quad 4 + 4 - \dfrac{4}{4}$

8 $\quad 4 + 4 + 4 - 4$

9 $\quad 4 + 4 + \dfrac{4}{4}$

10 $\quad 4 + 4 + \dfrac{4}{\sqrt{4}}$

11 $\quad \dfrac{4!}{\sqrt{4}} - \dfrac{4}{4}$

12 $\quad \dfrac{4!4}{4\sqrt{4}}$

13 $\quad \dfrac{4!}{\sqrt{4}} + \dfrac{4}{4}$

14 $\quad 4 + 4 + 4 + \sqrt{4}$

15 $\quad 4 \cdot 4 - \dfrac{4}{4}$

16 $\quad \dfrac{4 \cdot 4 \cdot 4}{4}$

17 $\quad 4 \cdot 4 + \dfrac{4}{4}$

18 $\quad 4 \cdot 4 + 4 - \sqrt{4}$

19 $\quad 4! - 4 - \dfrac{4}{4}$

20 $\quad \dfrac{4!4}{4} - 4$

21 $\quad 4! - 4 + \dfrac{4}{4}$

22 $\quad 4 \cdot 4 + 4 + \sqrt{4}$

23 $\quad \dfrac{4!4 - 4}{4}$

24 $\quad 4 \cdot 4 + 4 + 4$

25 $\quad \dfrac{4!4 + 4}{4}$

26 $\quad \dfrac{4!4}{4} + \sqrt{4}$

27 $\quad 4! + 4 - \dfrac{4}{4}$

28 $\quad \dfrac{4!4}{4} + 4$

29 $\quad 4! + 4 + \dfrac{4}{4}$

30 $\quad 4 \cdot 4 \cdot \sqrt{4} - \sqrt{4}$

31 $\quad 4! + 4 + \sqrt{\dfrac{4}{.4}}$

32 $\quad \dfrac{4 \cdot 4 \cdot 4}{\sqrt{4}}$

33 $\quad \dfrac{\dot{.4}(4!) + 4}{\dot{.4}}$

34 $\quad \dfrac{.4(4!) + 4}{.4}$

35 $\quad \dfrac{4 \cdot 4 - \dot{.4}}{\dot{.4}}$

36 $\quad \dfrac{4!\sqrt{4}}{4} + 4!$

37 $\quad \dfrac{4 \cdot 4 + \dot{.4}}{\dot{.4}}$

38 $\quad \dfrac{4}{\dot{.4}} \cdot 4 + \sqrt{4}$

39 $\quad \dfrac{4 \cdot 4 - .4}{.4}$

40   $\dfrac{4!}{\sqrt{4}} + 4! + 4$

41   $\dfrac{4 \cdot 4 + .4}{.4}$

42   $\dfrac{4 \cdot 4}{.4} + \sqrt{4}$

43   $\dfrac{4! - 4}{.4} - \sqrt{4}$

44   $\dfrac{4(4! - \sqrt{4})}{\sqrt{4}}$

45   $4!\sqrt{4} - \sqrt{\dfrac{4}{.4}}$

46   $4!\sqrt{4} - \dfrac{4}{\sqrt{4}}$

47   $\dfrac{4! - 4}{.4} + \sqrt{4}$

48   $\dfrac{4!4\sqrt{4}}{4}$

49   $4!\sqrt{4} + \dfrac{4}{4}$

50   $4!\sqrt{4} + \dfrac{4}{\sqrt{4}}$

51   $4!\sqrt{4} + \sqrt{\dfrac{4}{.4}}$

52   $\dfrac{4(4! + \sqrt{4})}{\sqrt{4}}$

53   $\dfrac{4! - \sqrt{4}}{.4} - \sqrt{4}$

54   $\dfrac{4! - 4}{.4} + 4$

55   $\dfrac{4! - .4}{.4} - 4$

56   $\dfrac{4! - .4 \cdot 4}{.4}$

57   $\dfrac{4! + .4}{.4} - 4$

58   $\dfrac{4! - .4\sqrt{4}}{.4}$

59   $\dfrac{4!}{.4} - \dfrac{4}{4}$

60   $\dfrac{4!\sqrt{4}}{.4\sqrt{4}}$

61   $\dfrac{4!}{.4} + \dfrac{.4}{.4}$

62   $\dfrac{4! + .4\sqrt{4}}{.4}$

63   $\dfrac{4!}{.4} + \sqrt{\dfrac{4}{.4}}$

64   $\dfrac{4! + .4 \cdot 4}{.4}$

65   $\dfrac{4! + .4}{.4} + 4$

66   $\dfrac{4!}{.4} + \dfrac{4}{\sqrt{.4}}$

67   $\dfrac{4! + 4}{.4} + 4$

68   $\dfrac{4!}{.4} + 4\sqrt{4}$

69   $\dfrac{4! + 4 - .4}{.4}$

70   $\dfrac{4!}{.4} + \dfrac{4}{.4}$

71   $\dfrac{4! + 4 + .4}{.4}$

72 $\dfrac{4! + 4}{.4} + \sqrt{4}$ 86 $4!4 - \dfrac{4}{.4}$

73 $\dfrac{4!\sqrt{4} + \sqrt{.4}}{\sqrt{.4}}$ 87 $4!4 - \dfrac{4}{.\dot{4}}$

88 $4!4 - 4\sqrt{4}$

74 $4 + \dfrac{4! + 4}{.4}$

89 $\dfrac{4! + \sqrt{4}}{.4} + 4!$

75 $\dfrac{4!\sqrt{4} + \sqrt{4}}{\sqrt{.4}}$

90 $\dfrac{4 \cdot 4}{(.4)(.\dot{4})}$

76 $4 + 4! \sqrt{\dfrac{4}{.4}}$

91 $4!4 - \dfrac{\sqrt{4}}{.4}$

77 $\left(\dfrac{4}{.\dot{4}}\right)^{\sqrt{4}} - 4$

92 $\sqrt{4}(4!\sqrt{4} - \sqrt{4})$

78 $\dfrac{4! + 4!(.\dot{4})}{.\dot{4}}$

93 $4!4 - \sqrt{\dfrac{4}{.\dot{4}}}$

79 $\left(\dfrac{4}{.\dot{4}}\right)^{\sqrt{4}} - \sqrt{4}$

94 $4!4 - \dfrac{4}{\sqrt{4}}$

80 $\dfrac{4!}{.4} + 4! + \sqrt{4}$

95 $4!4 - \dfrac{4}{4}$

81 $\left(\dfrac{4}{.\dot{4}}\right)^{\frac{4}{\sqrt{4}}}$

96 $\dfrac{4!4\sqrt{4}}{\sqrt{4}}$

82 $\dfrac{4!}{.\dot{4}} + 4! + 4$

97 $4!4 + \dfrac{4}{4}$

83 $\left(\dfrac{4}{.\dot{4}}\right)^{\sqrt{4}} + \sqrt{4}$

98 $4!4 + \dfrac{4}{\sqrt{4}}$

84 $\dfrac{4! + 4!(.4)}{.4}$

99 $4!4 + \sqrt{\dfrac{4}{.\dot{4}}}$

85 $\left(\dfrac{4}{.4}\right)^{\sqrt{4}} + 4$ 100 $\left(\dfrac{4}{.\dot{4}}\right)^{\frac{4}{\sqrt{4}}}$

AN INEQUALITY

If $X + Y + Z = 1$, prove $XY + YZ + XZ < \frac{1}{2}$.

Solution

Squaring $X + Y + Z = 1$ we have $X^2 + Y^2 + Z^2 + 2XY + 2YZ + 2XZ = 1$ or $XY + YZ + XZ = \frac{1}{2} - \frac{(X^2 + Y^2 + Z^2)}{2}$. Even if X, Y, or Z should be negative their squares must be positive; hence the expression in the brackets must be positive.

A problem that became an international affair, this puzzle was contributed by a mathematician in Kent, England, "simplified" by a reader in Holland, solved in an alternate way by a gentleman in Canada, and elaborated upon by two gentlemen in Massachusetts.

From Bronxville, New York, came the above solution.

From a reader at the Lincoln Laboratory of M.I.T., we received the following:

> ... let X, Y, Z be the roots of $w^3 - w^2 + bw + c = 0$. Then $X + Y + Z = 1$, and $XY + YZ + ZX = b$. Hence, if your hypothesis is provable, there can be no equations of the above type with $b \geq \frac{1}{2}$. An interesting sidelight of what I guess your proof to be though, says that $w^3 + aw^2 + bw + c = 0$ has only one real root if $2b \geq 2^2$.

A Massachusetts reader caught us for not having a more rigorous statement by including "provided X, Y, and Z are real numbers." Our wrists were slapped by the following analysis:

> ... you published the following mathematical puzzle: If $X + Y + Z = 1$, prove that $XY + YZ + XZ < \frac{1}{2}$.
>
> A quick analysis of the following proof will show a flaw in this problem:
>
> Multiplying the original equation by X and by Y and by Z, we obtain:

$$X^2 + XY + XZ = X$$
$$XY + Y^2 + ZY = Y$$
$$XZ + YZ + Z^2 = Z$$

Adding: $\quad X^2 + Y^2 + Z^2 + 2XY + 2YZ + 2XZ$
$$= X + Y + Z = 1$$

Rearranging terms:

$$2XY + 2YZ + 2XZ = 1 - (X^2 + Y^2 + Z^2)$$

This means that for the original statement to be proved, $X^2 + Y^2 + Z^2 > 0$, which is obviously true provided that X, Y, and Z are real numbers. It is possible, however, that $X^2 + Y^2 + Z^2 \leq 0$, if the imaginary component of X, Y, and Z is larger than the real component, in which case the original statement is incorrect; the statement cannot be proved.

As a simple example: let $X = 1$, $Y = i$, $Z = -i$, where $i = \sqrt{-1}$.

Then: $X + Y + Z = 1 + i - i = 1$.

But: $XY + YZ + XZ = i - i^2 - i = 1$ which is not $< \frac{1}{3}$.

Therefore this example serves as an adequate counterexample, since one case is sufficient to disprove the rule.

I thought that you might be interested in this basic error in your recent puzzle, so that you might set the record straight for other readers who may be confused.

The gentleman assured us, in conclusion, that he was an "avid follower" of our puzzles. We felt better.

The last word was received from The Hague, Netherlands. Rather than simplifying the puzzle, our correspondent proved a stronger theorem:

Your problem was too easy. Simplifying it we propose to prove that $XY + XZ + YZ$ equals or is smaller than $\frac{1}{3}$. We substitute $\frac{1}{3} + a$ for x, etc., so that we can write

$$(\tfrac{1}{3} + a + \tfrac{1}{3} + b + \tfrac{1}{3} + c) = 1$$

in which $a + b + c$ of course equals zero. $XY + YZ + XZ$ will then be written:

$$\tfrac{1}{3} + \tfrac{2}{3}(a + b + c) + ab + ac + bc$$

It's easy to prove that $ab + ac + bc \leq 0$ and since $\frac{2}{3}(a + b + c) = 0$, $XY + XZ + YZ \leq \frac{1}{3}$.

VARIATIONS ON THE TRICOLOR

Using the French Tricolor as a model, how many flags are possible with five available colors if two adjacent stripes must not be colored the same?

For solutions to problems in this section, turn to page 41.

A TRANSPOSED EQUATION

The algebra teacher wrote on the blackboard a quadratic equation of the form $x^2 - Ax + B = 0$. In copying this a careless student erroneously transposed the two digits of B as well as the plus and minus signs. However, one of the roots was the same. What was this root? (Assume both A and B are integers.)

A LADDER PROBLEM

A cubic box with 1-foot edges is placed flat against a wall. A ladder $\sqrt{15}$ feet long is placed in such a way that it touches the wall as well as the free horizontal edge of the box. Find at what height the ladder touches the wall.

THE BEAUTY CONTEST

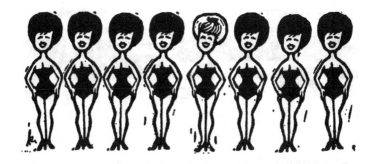

The beauty contest winner had a 36, 23, 34 figure. Although no two contestants had exactly the same measurements, the two runners-up differed by less than an inch in each measurement from the winner and the waist of each was two-thirds the hips of the other. If the sum of the three measurements was the same for all three girls, what were the vital statistics of the two runners-up? (The tape is accurate only to the quarter-inch.)

THE TWO MOTORISTS

Two motorists set out at the same time to go from A to B, a distance of 100 miles. They both followed the same route and traveled at different, though uniform, speeds of an integral number of miles per hour. The difference in their speeds was a prime number of miles per hour, and after they had been driving for 2 hours, the distance of the slower car from A was five times that of the faster car from B. How fast did the two motorists drive?

IMPROVING ON AN ESCALATOR

A certain physicist, who is always in a hurry, walks up an upgoing escalator at the rate of one step per second. Twenty steps bring him to the top. Next day he goes up at two steps per second, reaching the top in 32 steps. How many steps are there in the escalator?

AN AGE PROBLEM

Lottie and Lucy Hill are both 90 years old. Mary Jones, on the other hand, is half again as old as she was when she was half again as old as she was when she lacked 5 years of being half as old as she is now. How old is Mary?

THE FREIGHT TRAIN

Two men are walking toward each other alongside a railway. A freight train overtakes one of them in 20 seconds and exactly 10 minutes later meets the other man coming in the opposite direction. The train passes this man in 18 seconds. How long after the train has passed the second man will the two men meet? (Constant speeds are to be assumed throughout.)

A CLOSE RACE

Two hot rodders compete in a drag race. Each accelerates at a uniform rate from a standing start. Al covers the last quarter of the distance in 3 seconds; Bob covers the last third in 4 seconds. Who won?

THE WATCH CLOCKER

Dr. Reed, arriving late at the lab one morning, pulled out his watch and said, "I must have it seen to. I have noticed that the minute and the hour hand are exactly together every 65 minutes." Does Dr. Reed's watch gain or lose, and how much per hour?

THE EVASIVE ENGINEER

While visiting Cape Kennedy, we came upon an engineer digging a hole. "How deep is that hole?" we asked. "Guess," said the engineer, being evasive. "My height is exactly 5 feet 10 inches." "How much deeper are you going?" we inquired. "I am one-third done," was the answer, "and when I am finished my head will be twice as far below ground as it is now above ground." How deep will that hole be when finished?

A LIFETIME TASK

One of the largest known primes is $2^{3217} - 1$. Assume that it requires a human being a year to calculate each digit of this number. Who, if anyone, would have been capable of completing the job?

THE AGES OF MAN

A man passed one-sixth of his life in childhood, one-twelfth in youth, and one-seventh more as a bachelor. Five years after his marriage, a son was born who died four years before his father at half his father's final age. What was the man's final age?

AN INFINITE PRODUCT

For $|X| < 1$ evaluate the infinite product: $(1 + X + X^2 + \cdots + X^9)$ $(1 + X^{10} + X^{30} + \cdots X^{90})$ $(1 + X^{100} + X^{200} + \cdots + X^{900})$ $(\cdots)$

THE TARRY-TOWN RIDDLE

Between Sing-Sing and Tarry-Town
I met my worthy friend, John Brown,
And seven daughters, riding nags,
And every one had seven bags,
In every bag were thirty cats,
And every cat had forty rats,
Besides a brood of fifty kittens,
All *but* the nags and bags wore mittens!
Mittens, kittens—cats, rats—bags, nags—Browns,
How many were met between the towns?

THREE GENERATIONS

When I am as old as my father is now, I shall be five times as old as my son is now. By then my son will be eight years older than I am now. The combined ages of my father and myself are 100 years. How old is my son?

RENEWING OLD TIES

A railroad buys ties for $11 apiece. They last for 10 years and then have a scrap value of $1 apiece. If preservative treatment costing $3 a tie is applied, each tie will last 15 years but will have no scrap value. If the railroad makes 5 per cent on its capital, should it treat the ties?

ANOTHER AGE PROBLEM

A's age equals B's age plus the cube root of C's age. B's age equals C's age plus the cube root of A's age, plus 14 years. C's age equals the cube root of A's age plus the square root of B's age. What is the age of each?

A DIFFICULT FACTORIZATION

Express $X^{15} + 1$ as the product of sixth- and ninth-degree polynomials with integral coefficients.

THE COMPETING AIRLINES

There are nine cities which are served by two competing airlines. One or the other airline (but not both) has a flight between every pair of cities. What is the minimum number of possible triangular flights (i.e., trips from A to B to C and back to A on the same airline)?

THE OPTIMAL SALARY

Citizens of Franistan pay as much income tax (per-centage-wise) as they make rupees per week. What is the optimal salary in Franistan?

A TRIGONOMETRIC ROOT

Show that $\tan \dfrac{\pi}{10}$ is a root of the equation

$$5x^4 - 10x^2 + 1 = 0$$

THE BRICKLAYERS

A contractor estimated that one of his two bricklayers would take 9 hours to build a certain wall and the other 10 hours. However, he knew from experience that when they worked together, 10 fewer bricks got laid per hour. Since he was in a hurry, he put both men on the job and found it took exactly 5 hours to build the wall. How many bricks did it contain?

A STOCKMARKET PROBLEM

The stock of United Ticpolonga has been selling for the last year in the range of 70–80. A year ago the price of United stock was ¾ the price of its competitor, Jessurs, Inc. Jessurs is now selling at exactly ⅛ of a point below its price last year, but United has declined so far it is only ⅔ the price of Jessurs now. What are the two currently selling at?

THE FAMILY LEGACY

When Maharaja Ram Singh died, he left 3465 gold pieces to be divided equally among his children. Each wife had the same number of children and this number was 8 less than the number of wives per harem, which in turn was 4 more than the number of harems and 4 less than the number of gold pieces each child received. How many children did Ram Singh have?

A MATTER OF GREAT INTEREST

Obviously the smaller the compounding period, the greater the interest. How much does one dollar amount to after one year at 100 per cent per annum interest, compounded continuously, i.e., instantaneously?

THE SENIOR PROM

Tickets for the senior prom were $1.00 for boys and 65 cents for girls. Although there were more boys than girls at the dance, the percentage of boys who did not go was twice the percentage of girls who did not go. Knowing this percentage and the total senior class enrollment, one can deduce the total receipts for the affair. If this enrollment is between 60 and 100, what was the total attendance at the prom?

THE DIGITAL ATOM SMASHER

Does the square root of an ATOM extend from A to M?
Yes, if you can assign the proper numerical values to
the letters. $\sqrt{\text{ATOM}} = \text{A} + \text{TO} + \text{M}$. Here ATOM
is a four-digit number and TO is a two-digit number.

A SURE THING

On a surefire tip from Big Jim, Willy the Welcher placed some bets with Benny the Bookie. Big Jim had told Willy that in the third race at Holly Park one of the four outsiders was bound to win. Of the four, the first horse has odds of 3 to 1, the second 4 to 1, the third 5 to 1, and the fourth is 6 to 1. What must Willy bet on each horse to make a profit of $101, no matter which of the four outsiders wins?

TWO MEN ON A HORSE

Luke and Slim have only one horse between them. Luke rides a certain time and then ties up the horse for Slim, who has been walking. Meanwhile Luke walks on ahead. They proceed in this way, alternately walking and riding. If they walk 4 miles per hour and ride 12 miles per hour, what part of the time is the horse resting?

A DECREASING RATIO

John was three times as old as his sister 2 years ago and five times as old 2 years before that. In how many years will the ratio be 2 to 1?

AN INFINITUDE OF TWOS

Assuming that the expression $\sqrt{2 + \sqrt{2 + \sqrt{2 + \cdots}}}$ converges to a finite limit, evaluate it.

Solutions

Variations on the Tricolor

First there are 20 flags whose top and bottom rows are the same color (5 choices for the middle row, then 4 choices for the outside rows). Then there are $5 \times 4 \times 3$ or 60 flags containing 3 colors. But half of these can be obtained by turning the other half upside down, so the actual total is $20 + 30$ or 50 flags.

A Transposed Equation

The equation must have been either $x^2 - 11x + 24 = 0$ (which was copied as $x^2 + 11x - 42 = 0$) or $x^2 - 22x + 57 = 0$ (which was copied as $x^2 + 22x - 75 = 0$). In either case the common root is 3.

A Ladder Problem

Let the legs of the right triangle formed be x and y.

Then: $$x^2 + y^2 = 15 \quad \text{and} \quad \frac{x-1}{1} = \frac{1}{y-1}$$

or: $$xy = x + y$$

Then $x^2 + y^2 + 2xy = (x+y)^2 = 2(x+y) + 15$. Solving as a quadratic in $x + y$, we have $x + y = 5$ or -3. Rejecting the negative root, $x + y = 5$, whence $x^2 - 5x + 5 = 0$. The two roots of this quadratic, 3.62 and 1.38, represent, in feet, the two possible solutions.

The Beauty Contest

Let the measurements of the two runners-up be a, b, c; x, y, z, respectively. Then $a + b + c = x + y + z = 93$, $b = \frac{2}{3}z$, and $y = \frac{2}{3}c$. Because of the $\frac{1}{4}$-inch accuracy restriction, $4c$ and $4z$ must both be divisible by 3. The only possibilities are $33\frac{1}{4}$ and $34\frac{1}{2}$, and to avoid duplication of the entire set of measurements among the runners-up, one must have a hip measurement of $33\frac{1}{4}$, the other $34\frac{1}{2}$. The other measurements are readily obtained, and we get 36, $22\frac{1}{2}$, $34\frac{1}{2}$ and $36\frac{1}{4}$, 23, $33\frac{3}{4}$.

The Two Motorists

Let $d =$ the distance the faster car has traveled. Then it still has $100 - d$ miles to go, and the slow car has, therefore, gone $500 - 5d$ miles. Let f and s be the rates of the fast and slow cars in mph. Then

$$\frac{d}{f} = \frac{500 - 5d}{s} = 2.$$ Solving, we get $s = 250 - 5f$. Then $f - s = 6f - 250 = 2(3f - 125)$. But since $f - s$ is prime $3f - 125 = 1$. Hence $f = 42$ mph and $s = 40$ mph.

Improving on an Escalator

Let $s =$ the number of steps and $r =$ the rate of the escalator (steps per second). Then:

$$s - 20 = 20r$$
$$\text{and} \quad s - 32 = 16r$$

Solving simultaneously, $s = 80$.

An Age Problem

The equation $2/3 \cdot 2M/3 + 5 = \frac{1}{2}M$ gives the solution $M = 90$. In other words, Mary is as old as the Hills.

The Freight Train

Let $L =$ length of the train
Let $A =$ distance first man walks in 20 seconds
Let $B =$ distance second man walks in 18 seconds

Then the train goes $L + A$ in 20 seconds and $L - B$ in 18 seconds. Hence the train travels in 10 minutes a distance of:

$$30(L + A) = \frac{100(L - B)}{3} \quad \text{or} \quad L = 9A + 10B$$

In 10 minutes A travels a distance of $30A$. Hence the distance separating A and B at the moment the train reaches B is:

$$30(L + A) + L - 30A = 31L$$

Substituting $9A + 10B$ for L, we get $31(9A + 10B)$ as the distance separating A and B at the moment the train reaches B. Now if the first man takes 20 seconds to walk A and the second man takes 18 seconds to walk B, then they reduce the distance between themselves by $9A + 10B$ in 180 seconds. Therefore, it will take them

$$\frac{31(9A + 10B) \cdot 180}{9A + 10B}$$

seconds to cover the distance between them. From this must be deducted 18 seconds which it takes the train to pass B. Thus the required time is 5562 seconds = 1 hour, 32 minutes, 42 seconds.

42

A Close Race

Al covers the last $\frac{3}{12}$ in 3 seconds; Bob, the last $\frac{4}{12}$ in 4 seconds. Both average $\frac{1}{12}$ of the course per second for the respective distances. Since Bob maintains this average rate over a greater distance, he must achieve it first. Hence he is the winner.

The Watch Clocker

It should take the hands $\frac{12}{11}$ hours to come together, not 65 minutes or $\frac{13}{12}$ hours as it actually does. Therefore, Dr. Reed's watch runs too fast by a factor of $(12/11 \div 13/12) = \frac{144}{143}$. In one hour it gains $[(144/143) - 1]$ hours $= 60/143$ minutes $\doteq 25$ seconds.

The Evasive Engineer

When completed the hole will be 10 feet 6 inches deep. $x =$ depth now, $70 - x =$ height above now, $3x =$ finished depth $= 70 + 2(70 - x) \therefore 3x = 126 = 10$ feet 6 inches.

A Lifetime Task

Only one man, Methuselah, who attained the record age of 969 years. Solving the equation $2^{3217} = 10^x$ we have $x = 3217 \log 2 = 968.4$. The number of digits in $2^{3217} - 1$ is one more than the integral part of x or 969.

The Ages of Man

Let $x =$ man's final age. Then

$$x = x/6 + x/12 + x/7 + 5 + x/2 + 4$$

with solution $x = 84$.

An Infinite Product

Since every integer has a unique decimal representation, every non-negative integral power of x will appear exactly once in the product. Hence the product equals:

$$\sum_{n=0}^{\infty} x^n = 1/(1 - x)$$

The Tarry-Town Riddle

8 Browns; 8 nags; 56 bags; 1,680 cats; 67,200 rats; 84,000 kittens and 611,536 mittens — a total of 764,488.

Three Generations

Let M = my age in years, S = son's age, and F = father's age. My father is five times as old as my son, therefore: $F = 5S$. When I am as old as my father, my son will be $F - M$ years older than he is now; his age then will be $S + F - M$. Since this is 8 years more than my present age, $S + F - M = M + 8$. The third equation is: $F + M = 100$. Solving these equations gives $S = 13$, and hence $M = 35$, and $F = 65$.

Renewing Old Ties

Each tie represents an outlay of $11 + 10v^{10} + 10v^{20} \cdots$ where $v = \dfrac{1}{1.05}$ or \$26.90. If the ties are treated the corresponding outlay per tie will be $14 + 14v^{15} + 14v^{30} + \cdots$ or \$26.98. Therefore, it is more economical not to treat the ties.

Another Age Problem

Denoting by X^3, Y^2, Z^3, respectively the ages of A, B, C, we have the equations: $X^3 = Y^2 + Z$; $Y^2 = Z^3 + X + 14$; $Z^3 = X + Y$. From these it is possible to obtain an equation of the 18th degree, but the only rational values of X, Y, and Z can, without too much difficulty, be obtained from the original equations. $X = 3$, $Y = 5$, $Z = 2$ and A is 27 years old, B is 25, C is 8.

A Difficult Factorization

$X^{15} + 1 = (X^3)^5 + 1 = (X^3 + 1)P(X) = (X + 1)(X^2 - X + 1)P(X)$ where $P(X)$ is a 12th-degree polynomial. Also $X^{15} + 1 = (X^5)^3 + 1 = (X^5 + 1)Q(X) = (X + 1)(X^4 - X^3 + X^2 - X + 1)Q(X)$ where $Q(X)$ is a 10th-degree polynomial. Since the roots of $X^{15} + 1$ are the 15 fifteenth roots of -1 (all distinct), the product

$$(X^2 - X + 1)(X^4 - X^3 + X^2 - X + 1)$$
$$= X^6 - 2X^5 + 3X^4 - 3X^3 + 3X^2 - 2X + 1$$

must be a factor of $X^{15} + 1$. The 9th-degree factor can be obtained by division.

The Competing Airlines

If the number of cities is of the form $4n + 1$, the minimum number of triangular flights is $\binom{2n}{3} + \binom{2n + 1}{3} - n$. In the problem $n = 2$.

Therefore $\binom{4}{3} + \binom{5}{3} - 2$ or $4 + 10 - 2 = 12$ is the number of triangular flights. This minimum number can be obtained from many different arrangements. One matrix corresponding to a complete nonagon might look like this:

	1	2	3	4	5	6	7	8	9
1		A	B	A	A	A	B	B	B
2			B	B	A	B	A	B	A
3				A	A	B	B	A	A
4					B	B	A	A	B
5						A	B	B	B
6							B	A	A
7								A	A
8									B

Here the cities are numbered from 1 to 9 and the two airlines are A and B, the table showing the airline serving any two cities. There are exactly 12 triangles which are either all A or all B.

The Optimal Salary

On a salary of $50 \pm x$ rupees, a Franistanian pays

$$\left(\frac{50 \pm x}{100}\right) \cdot (50 \pm x)$$

rupees and is left with $25 - \dfrac{x^2}{100}$ rupees. Since $x = 0$ maximizes the take-home pay, 50 rupees is the optimal salary.

A Trigonometric Root

The angle $a = \dfrac{\pi}{10}$ satisfies the relation $\tan 3a = \cot 2a$, i.e.,

$$\frac{3 \tan a - \tan^3 a}{1 - 3 \tan^2 a} = \frac{1 - \tan^2 a}{2 \tan a}$$

or $5x^4 - 10x^2 + 1 = 0$, where $x = \tan a$.

The Bricklayers

Let N = number of bricks in wall, $\dfrac{N}{9}$ = number of bricks first brick-layer lays per hour, $\dfrac{N}{10}$ = number of bricks second bricklayer lays per hour, $\dfrac{N}{9} + \dfrac{N}{10} - 10 =$ number of bricks laid per hour when they work

together, and finally $\dfrac{N}{\dfrac{N}{9} + \dfrac{N}{10} - 10} = 5$, from which $N = 900$.

A Stockmarket Problem

If we let U_0 and U_1 be the price of United stock a year ago and today, we have $3/2U_1 = 4/3U_0 - \frac{1}{8}$. U_0 must be equal to or less than 80, U_1 must be equal to or greater than 70. Since stockmarket prices are only quoted to eighths, $U_0 = 79\frac{1}{8}$, $U_1 = 70\frac{1}{4}$ is the only solution. The present prices are, therefore: United $70\frac{1}{4}$, Jessurs $105\frac{3}{8}$.

The Family Legacy

Let number of children $= x$ per wife; then number of harems $= x + 4$, number of wives per harem $= x + 8$, and number of gold pieces per child $= x + 12$. Then

$$x(x + 4)(x + 8)(x + 12) = 3465$$

$$\frac{x}{4}\left(\frac{x}{4} + 1\right)\left(\frac{x}{4} + 2\right)\left(\frac{x}{4} + 3\right) = \frac{3465}{256}$$

Factoring, we get:

$$\left(\frac{x^2 + 12x + 16}{16}\right) - 1 = \frac{3465}{256}$$

$$(x^2 + 12x + 16)^2 - 256 = 3465$$
$$x^2 + 12x + 16 = 61$$
$$x_1 = -15 \qquad x_2 = +3$$

Rejecting the negative root, $x = 3$, and the number of children is 231.

A Matter of Great Interest

$A = \lim\limits_{t \to 0} (1 + t)^{1/t}$ where t is the compounding period in years. Thus $A = \$e = \$2.71^+$.

The Senior Prom

Let $N =$ total number in senior class, of which B are boys and let $p =$ proportion of girls who did not go to the dance. We have then: Total receipts in dollars $= B(1 - 2p) + 0.65(N - B)(1 - p)$ or $0.35B - 1.35Bp + 0.65N - 0.65Np$. If one can deduce the total receipts from a knowledge of p and N, this expression must be independent of B; therefore $p = \frac{7}{27}$. Since the total enrollment is between 60 and 100, $B = 54$ and the number of girls $= 27$, so that there were 26 boys and 20 girls or 46 altogether at the dance.

The Digital Atom Smasher

$$\sqrt{1296} = 1 + 29 + 6 \quad \text{or} \quad \sqrt{6724} = 6 + 72 + 4$$

A Sure Thing

Willy bets 105, 84, 70, and 60 dollars on each horse with the respective odds 3, 4, 5, and 6 to 1. Willy has to make back the money bet on each horse plus 101 dollars. Therefore:

$$4W = 5X = 6Y = 7Z = W + X + Y + Z + 101$$

(four equations with four unknowns)

From these equations we have: $W = 105$, $X = 84$, $Y = 70$, and $Z = 60$, which are the amounts bet on each horse with the respective odds 3, 4, 5, 6 to 1. The reason one can make a profit regardless of the winning horse is that $\frac{1}{4} + \frac{1}{5} + \frac{1}{6} + \frac{1}{7} < 1$.

Two Men on a Horse

If each man walks D miles and then rides D miles in time T,

$$\frac{D}{4} + \frac{D}{12} = T, \text{ from which } D = 3T$$

Since they cover $6T$ miles in T hours, ($3T$ walking and $3T$ riding), they progress 6 miles an hour. The horse's speed is 12 mph; therefore, he rests half the time.

A Decreasing Ratio

Let $x =$ the age of the sister 4 years ago. Then $5x + 2 = 3(x + 2)$, giving $x = 2$. Hence John is 14 and his sister is 6. In two years the ratio will be 2 to 1.

An Infinitude of Twos

Let $x = \sqrt{2 + \sqrt{2 + \sqrt{2 +}}} \cdots$. Squaring both sides, we have $x^2 = 2 + \sqrt{2 + \sqrt{2 +}} \cdots = 2 + x$ or $x^2 - x - 2 = 0$ with roots $x = -1, 2$. Rejecting the negative root, we conclude that

$$\sqrt{2 + \sqrt{2 + \sqrt{2 +}}} \cdots = 2$$

2

AXIOMS, ANGLES, AND ARCS

Geometric Exercises

The Shortest Ladder (SOLUTION, Page 52)

A Noncompass Construction (SOLUTION, Page 53)

Very Similar Triangles · *The Bridges* · *Separating the Sheep* · *The Atom Smasher* · *A Polyhedron Problem* · *A Curious Sphere* · *Halving a Triangle* · *A Max-min Problem* · *The Pie-shaped Field* · *A Plethora of Circles* · *The Clockwatchers* · *A Sphere-packing Problem* · *Changing the Base*

SOLUTIONS: Page 68

THE SHORTEST LADDER

A ladder is leaning against a wall at an angle steeper than 45 degrees. Under the ladder there is a barrel which touches both the ladder and the wall. If the vertical distance, in feet, between the top of the ladder and the ground is four times the diameter of the barrel, what is the shortest integral number of feet the ladder can be?

Solution

Let z be the length of the ladder. Then $x^2 + y^2 = z^2$, where x and y are the segments of wall and floor determined by the ladder. The parametric solution of this well-known equation is $x = a^2 - b^2$; $y = 2ab$, $z = a^2 + b^2$. It will also be found that the diameter of the largest inscribable circle is given by $2b(a - b)$. A little trial and error soon reveals that in this case $a = 4$ and $b = 3$, yielding $x = 7$; $y = 24$ and $z = 25$.

This was one of our fiascos. Neither the wording of the problem nor the accompanying illustration give a clue as to the assumed orientation of the barrel. The originator meant for the barrel to be lying on its side with its circular end facing the reader, but solvers throughout the country saw it differently. Most of them either had the barrel standing upright or lying with its circular ends against the wall and the ladder. Some chose to orient the barrel less symmetrically. Alternate solutions ranged from 15 feet downward and, in general, involved some of the weirdest-shaped barrels you ever saw.

A reader in Illinois made the following observation after the publication of the original edition of this book: "The problem is still a fiasco. The problem asks what is the shortest length of the ladder if restricted to integers. No restriction is placed on sides x and y. For $a = 4/5$ and $b = 3/5$, $z = 1$, while $x = 7/25$ and $y = 24/25$."

We print the problem both as a reminder to ourselves that a clear, unambiguous problem statement is essential and as a monument to the ingenuity (and imagination) of our readers.

A NONCOMPASS CONSTRUCTION

From a point P outside a given circle, construct, using a straight edge only, the perpendicular to a given diameter of the circle.

Solution

Let A and B be the extremities of the diameter. Construct the line through P and A intersecting the circle at C, and let D be the corresponding point on the circle by considering the line through P and B. The intersection of the lines through A and D, B and C, yields a point E with the line through P and E perpendicular to the line through A and B.

Our proof is based on two theorems from high school geometry: (1) angles inscribed in semicircles are right angles and (2) the altitudes of a triangle meet in a common point. We received some alternate solutions such as this one from a reader at Norton Air Force Base:

> Let A and B be the extremities of the diameter. Construct tangents to the circle through point P, the tangential points being C and D. The intersection of lines CB and AD yields a point E, and the line through P and E is perpendicular to the line through A and B.

The contributors forgot, however, the condition that only a straight edge was to be used in the construction. Obtaining the points of tangency would require the use of compasses.

We had to pause and ponder when a solver from Houston, Texas, asked ". . . and if P is located like below, then what?"

Did our solution require that triangle APB be acute? A card from E. D. Friedman of Plainview, New York, relieved our doubts. Far from holding true only for limited positions of P, Mr. Friedman pointed out that we had been unnecessarily restrictive in our statement of the problem and demonstrated that the published construction was valid for all points P outside the circle and for all points P inside the circle as well!

VERY SIMILAR TRIANGLES

Two similar triangles with integral sides have two of their sides the same. The third sides differ by 387. What are the lengths of the sides?

For solutions to problems in this section, turn to page 68.

THE BRIDGES

A divided highway goes under a number of bridges, the arch over each lane being in the form of a semi-ellipse with the height equal to the width. A truck is 6 feet wide and 12 feet high. What is the lowest bridge under which it can pass?

SEPARATING THE SHEEP

A 1-acre field in the shape of a right triangle has a post at the midpoint of each side. A sheep is tethered to each of the side posts and a goat to the post on the hypotenuse. The ropes are just long enough to let each animal reach the two adjacent vertices. What is the total area the two sheep have to themselves, i.e., the area the goat cannot reach?

THE ATOM SMASHER

A new kind of atom smasher is to be composed of two tangents and a circular arc which is concave toward the point of intersection of the two tangents. Each tangent and the arc of the circle is 1 mile long. What is the radius of the circle?

A POLYHEDRON PROBLEM

The faces of a solid figure are all triangles. The figure
has nine vertices. At each of six of these vertices, four
faces meet, and at each of the other three vertices, six
faces meet. How many faces does the figure have?

A CURIOUS SPHERE

The area and volume of a certain sphere are both four-digit integers times π. What is the radius of the sphere?

HALVING A TRIANGLE

Given a point P on one side of a general triangle ABC, construct a line through P which will divide the area of the triangle into two equal halves.

A MAX-MIN PROBLEM

Given five points in or on a unit square, prove that at least two points are no farther than $\dfrac{\sqrt{2}}{2}$ units apart.

THE PIE-SHAPED FIELD

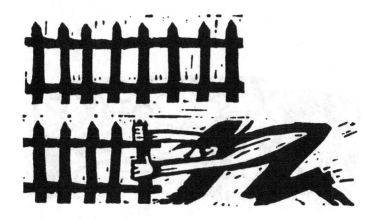

A farmer owned a square field measuring exactly 2261 yards on each side. 1898 yards from one corner and 1009 yards from an adjacent corner stood a beech tree. A neighbor offered to purchase a triangular portion of the field, stipulating that a fence should be erected in a straight line from one side of the field to an adjacent side so that the beech tree was part of the fence. The farmer accepted the offer but made sure that the triangular portion was of minimum area. What was the area of the field the neighbor received, and how long was the fence?

A PLETHORA OF CIRCLES

A circle of radius 1 inch is inscribed in an equilateral triangle. A smaller circle is inscribed at each vertex, tangent to the circle and to two sides of the triangle. The process is continued with progressively smaller circles. What is the sum of the circumference of all the circles?

THE CLOCKWATCHERS

A clock hangs on the wall of an Early Warning Display and Control Center, 71 feet 9 inches long and 10 feet 4 inches high. (Those are the dimensions of the wall, not of the clock!) While waiting for the waning crescent moon to rise, we noticed that the hands of the clock were pointing in opposite directions and were parallel to one of the diagonals of the wall. What was the exact time?

A SPHERE-PACKING PROBLEM

A man packing 1-inch spheres into a rectangular tray fills the tray in a single layer with no slack, using a rectangular packing. Trying a different arrangement, he fits in one more sphere. He then uses a third arrangement and fits in still another sphere. What is the size of the tray?

CHANGING THE BASE

An isoceles triangle has a 10-inch base and two 13-inch sides. What other value can the base have and still yield a triangle with the same area?

Solutions

Very Similar Triangles

In order for two similar triangles to have two sides the same, the sides must be of the form pm^3, pm^2n, and pmn^2 for the first triangle and pm^2n, pmn^2, and pn^3 for the corresponding sides of the second. The difference between the two sides not common to the two triangles is, therefore, $p(m^3 - n^3)$. The only solution of $p(m^3 - n^3) = 387$ in positive integers is $p = 1$, $m = 8$, $n = 5$. The triangles, therefore, have sides of 512, 320, 200, and 320, 200, 125.

The Bridges

We have

$$\frac{x^2}{a^2} + \frac{y^2}{4a^2} = 1$$

as the equation of the ellipse, where $2a$ is the height of the arch. From the data, $x = 3$, $y = 12$, so that $a = 3\sqrt{5}$. The truck will therefore be able to go under an arch $6\sqrt{5}$ or approximately 13 feet 5 inches high.

Separating the Sheep

The sheep have to themselves the two crescents cut from the semicircles on the sides by the semicircle on the hypotenuse. The area of these crescents is equal to the sum of the triangle and the two smaller semicircles minus the area of the semicircle on the hypotenuse. Since the sum of the two smaller semicircles is equal to the area of the semicircle on the hypotenuse, the area of the two crescents is equal to the area of the triangle; the two sheep have exactly 1 acre to themselves.

The Atom Smasher

Let AT and BT be the tangents, ACB the arc, O the center, OA the radius. Let angle $AOT = x$ and $AO = r$. Then

$$\text{arc } ACB = (2\pi - 2x)r = 1 \quad\text{and}\quad \tan x = 1/r$$

whence $\tan x = 2(\pi - x)$. By approximation: $x = 74° \, 46.2'$. $r = 1$ mile/$\tan x = 5280$ feet/$\tan x = 1437.45$ feet.

A Polyhedron Problem

There are 21 edges. A famous formula of Euler says vertices − edges + faces = 2 for any polyhedron, so that the number of faces is 14.

A Curious Sphere

Both $4R^2$ and $\frac{4}{3}R^3$ must lie between 1000 and 9999. From the first condition, $50 > R > 15$ and from the second, $20 > R > 9$. Therefore, R lies between 16 and 19 inclusive. But for $\frac{4}{3}R^3$ to be an integer, R must be divisible by 3. Therefore $R = 18$, area $= 1296\,\pi$, volume $= 7776\,\pi$.

Halving a Triangle

From P draw a line, L, to the opposite vertex, say A. Now construct a line parallel to L from the midpoint of BC, intersecting the side of the triangle at Q. The line PQ divides the triangle into two equal areas.

A Max-min Problem

Draw two lines through the center of the square, perpendicular to each other, such that each is parallel to a side of the unit square. These two lines partition the unit square into four $\frac{1}{2}$-unit squares. At least 2 of the 5 points must be in (or on the perimeter of) one of these smaller squares. This pair of points cannot be farther apart than the length of the small square's diagonal, $\dfrac{\sqrt{2}}{2}$ units.

The Pie-shaped Field

The area of the triangle will be minimal if the beech tree bisects the side on which it lies. The area of the triangular portion is 939,120 square yards. The length of the fence is 2018 yards.

A Plethora of Circles

Determine the height of the equilateral triangle to be 3 inches. The sum of the diameters of the circles at each vertex is 1 inch, and the sum of the diameters of all circles is 5 inches. Since $\sum \pi d_i = \pi \sum d_i$ the sum of the circumferences is 5π inches.

The Clockwatchers

With the aid of trigonometry tables and the most basic knowledge involving "possible" and "impossible" times, the solution of $43\frac{7}{11}$ minutes after 2 A.M. is determined.

A Sphere-packing Problem

The tray is 8 inches by 6 inches. Rectilinearly packed it holds 48 spheres. With the centers placed 60 degrees to each other, a ninth column can be obtained. With 5 spheres in the first column and 6 spheres in the second one and then alternating, the tray holds 49. With 6 spheres in the first column and 5 in the second, etc., the tray holds 50.

Changing the Base

Two 5-inch $\times$ 12-inch $\times$ 13-inch right triangles can be put together in two ways to form an isoceles triangle with equal 13-inch sides. One way involves a base of 10 inches, the other 24 inches. Naturally the area is the same in either case.

3

SOLVING IN INTEGERS

1234567891234567891234567891234567891234567891234567891234567891 2345

Diophantine Diversions

A Digit Problem (SOLUTION, Page 74)

The Sum of Two Cubes (SOLUTION, Page 76)

The Two Fences · The Absent-minded Student · The Two Cubes · Byzantine Basketball · The Two Sails · A Mathematical Impossibility · The Road Network · The Reciprocal Pythagorean Equation · A Double Coincidence · A Diophantine Equation · Finding the Coefficients · Another Diophantine Equation

SOLUTIONS; Page 86

A DIGIT PROBLEM

Find a three-digit number that is the sum of the cubes of its digits.

Solution

$$153 = 1^3 + 5^3 + 3^3 \quad \text{or} \quad 371 = 3^3 + 7^3 + 1^3$$

We spoke of our ability to err in our Introduction. This Diophantine number theory problem is a good case in point. Our two published answers were not the *only* answers as a number of our readers were quick to point out.

An Air Force Lieutenant in California neatly summed up our oversight as follows: "Webster defines a 'digit' as 'any of the figures 1 to 9 inclusive and, usually, the symbol 0.' "

No restriction against zero was posed in the statement of the problem. Thus the two additional answers submitted by our readers are correct:

$$370 = 3^3 + 7^3 + 0^3 \quad \text{and} \quad 407 = 4^3 + 0^3 + 7^3$$

From a reader in Baltimore, Maryland, came this concise statement:

> The solution to #140, a number equal to the sum of the cubes of its digits, gave only two of the four answers. The complete answer is 153, 370, 371, and 407. This can be generalized by stating that there are no answers for two-digit numbers, no answers for four-digit numbers, and none for numbers of greater length. For one digit, there are trivial solutions of 0 and 1. Thus, the problem could be given without the restriction that the number is three digits long.

A gentleman from RCA in New Jersey wrote:

> May I say first that in general I thoroughly enjoy your Problematical Recreations, as do many others among us here at RCA. However, I would like to take the liberty of offering some criticism, of the constructive kind only, with the aim of improving the quality and enjoyment of the problems for those of us concerned with precision and rigor. May I suggest first that you publish *all* the solutions to your problems. . . .

The best we could promise was to furnish all solutions we could possibly find.

74

THE SUM OF TWO CUBES

What is the smallest number that is the sum of two cubes in two different ways?

Solution

$$12^3 + 1^3 = 10^3 + 9^3 = 1729$$

We thought our solution was the smallest number. And so it is if you assume positive integers have been implied. However, the problem is not so stated, and even in the original, credited to Ramanujan, it is not precisely expressed. Clearly another case of insufficiently stated conditions.

The following answer, considerably smaller than ours, from an astute reader in San Bernardino, California, using one negative cube cannot be ruled out:

> The following solution which is $10^3 + 1^3$ less than yours is submitted for your consideration:
>
> $$8^3 + 6^3 = 9^3 + (-1)^3 = 728$$

It occurred to us, however, that we could go our astute reader one better with the solution: $(-12)^3 + (-1)^3 = (-10)^3 + (-9)^3 = -1729$. Carrying on in this fashion, one can obtain as small a solution as desired.

THE TWO FENCES

A farmer used 139 yards of fencing to enclose a rectangular field and to construct a fence along one of the diagonals of length 41 yards. He then found that a neighbor had fenced a one-third larger rectangular area in the same manner with less fencing. If all dimensions are integral yards, what are the dimensions of the neighbor's field?

For solutions to problems in this section, turn to page 86.

THE ABSENT-MINDED STUDENT

A brilliant graduate mathematics student was working on an assignment but, being a bit absent-minded, he forgot whether he was to add or to multiply the three different integers on his paper. He decided to do it both ways and, much to his surprise, the answer was the same. What were the three different integers?

THE TWO CUBES

Two cubes with integral sides have their combined volume equal to the combined length of their edges. What are the dimensions of the cubes?

BYZANTINE BASKETBALL

In Byzantine basketball there are 35 scores which are impossible for a team to total, one of them being 58. Naturally a free throw is worth fewer points than a field goal. What is the point value of each?

THE TWO SAILS

At the local yachting club they were checking over their equipment in preparation for the coming season. Woodson, a mathematically minded member, remarked that the two sails on his boat shared a curious property: the area of each in yards was equal to its perimeter in yards. Both sails, it should be mentioned, were of the usual right triangular shape and were of different dimensions, all dimensions being integers. What must have been their areas?

A MATHEMATICAL IMPOSSIBILITY

Prove that the product of four consecutive positive integers cannot be a perfect square.

THE ROAD NETWORK

A, *B*, and *C* are three towns, each pair being connected by a network of roads. A motorist notices that there are 82 routes from *A* to *B*, including those via *C* and 62 routes from *B* to *C*, including those via *A*. He also notices that there are fewer than 300 routes from *A* to *C*, including those via *B*. How many are there?

THE RECIPROCAL PYTHAGOREAN EQUATION

Find the smallest solution in integers of the equation
$$\frac{1}{x^2} + \frac{1}{y^2} = \frac{1}{z^2}.$$

A DOUBLE COINCIDENCE

Smith said to Jones, "I just bought four mujibs at $21.78 apiece, and I noted a curious thing. The total was $87.12, the price of a mujib in reverse order." "Isn't that a coincidence?" said Jones. "The other day I bought some glinches (no, not one or four) and I remarked the same thing." How much does a glinch cost and how many did Jones buy?

A DIOPHANTINE EQUATION

Solve in integers x and y the equation $x^2 = \dfrac{y^2}{y+4}$.

FINDING THE COEFFICIENTS

Find positive integers A and B $(A < B)$ such that both of the quadratics $x^2 + Ax + B$ and $x^2 + Bx + A$ factor into integers.

ANOTHER DIOPHANTINE EQUATION

Solve $x^2 + 2 = y(x+3)$ in positive integers x and y.

Solutions

The Two Fences

Let the dimensions of the first field be a and b. Then by the well-known parametric formula

$$a = 2rmn$$
$$b = r(m^2 - n^2)$$
$$41 = r(m^2 + n^2)$$

Since 41 is prime, $r = 1$. Also 41 can be expressed as the sum of two squares in just one way, $25 + 16$. Hence $m = 5$, $n = 4$, $b = 9$, and $a = 40$. The area is then 360 square yards. The neighbor's field, therefore, has area = 480 square yards. The only factorization of 480 which gives two legs of a Pythagorean triangle is $(16)(30)$, which represents the dimensions in yards of the neighbor's field. The diagonal of the neighbor's field is, therefore, 34 yards and his total fencing only 126 yards. It will be noticed that the condition that the first farmer used 139 yards of fencing is superfluous.

The Absent-minded Student

One, two, and three. We have $x + y + z = xyz$, so

$$\frac{1}{yz} + \frac{1}{xz} + \frac{1}{xy} = 1$$

If $x \leq y \leq z$ then

$$\frac{1}{xy} \geq \frac{1}{yz} \quad \text{and} \quad \frac{1}{xz}, \text{ so } \frac{3}{xy} \geq 1$$

and $xy \leq 3$. Hence $x = 1$ and $y = 1$, 2, or 3. But only the value $y = 2$ satisfies the original equation, giving $z = 3$.

The Two Cubes

$x^3 + y^3 = 12(x + y)$ or $x^2 - xy + y^2 = 12$, $x^2 - 2xy + y^2 = 12 - xy$, $(x - y)^2 = 12 - xy \geq 0$. If $x \leq y$ then $12 \geq x^2$ and $x = 1$, 2, or 3. The only integral solution is $x = 2$, $y = 4$.

Byzantine Basketball

The point values must be coprime; otherwise there are infinitely many impossible scores. The number of impossible scores is then: $1/2(x - 1)(y - 1) = 35$. Hence x and y are (2, 71), (3, 36), (6, 15), or (8, 11). (3, 36) and (6, 15) are eliminated by the coprimality condition, and since a score of 58 could be attained by making 29 two-point free throws, the value of a free throw is 8 and that of a field goal is 11.

The Two Sails

Let x, y and z be the legs and hypotenuse of a right triangle whose area and perimeter are equal. By the well-known parametric formula for Pythagorean triples,

$$x = 2abr$$
$$y = r(a^2 - b^2)$$
$$z = r(a^2 + b^2)$$

Then $abr^2(a^2 - b^2) = 2ar(a + b)$ or $br(a - b) = 2$. Thus b is at most 2. If $b = 2$, $r(a - 2) = 1$, giving $r = 1$, $a = 3$. Otherwise $b = 1$, in which case $r(a - 1) = 2$, giving two solutions: $r = 1$, $a = 3$ or $r = 2$, $a = 2$. The solution $(a, b, r) = (3, 2, 1)$ gives a sail of dimensions 5, 12, and 13 yards with area $= 30$ square yards. The last two solutions give the same dimensions of 6, 8, and 10 yards with area $= 24$ square yards.

A Mathematical Impossibility

Let N be the smallest integer. The product is then

$$N(N + 1)(N + 2)(N + 3) = (N^2 + 3N)(N^2 + 3N + 2)$$
$$= (N^2 + 3N + 1)^2 - 1$$

This is not a perfect square since two positive squares cannot differ by 1.

The Road Network

46 different routes. Letting x, y and z represent respectively the number of direct routes from A to B, B to C, and C to A, we have the following equations: $x + yz = 82$, $y + xz = 62$. These two equations give:

$$y = \frac{2(31 - 41z)}{(1 - z)(1 + z)}$$

which gives rise to three values of z: 2, 3, and 11 with corresponding solutions: 478, 302, and 46. Only the latter meets the conditions of the problem.

The Reciprocal Pythagorean Equation

The smallest solution is where $x = 15$, $y = 20$, $z = 12$. The general formula is: $x = m^4 - n^4$; $y = 2mn(m^2 + n^2)$; $z = 2mn(m^2 - n^2)$.

A Double Coincidence

$n(1000x + 100y + 10u + v) = 1000v + 100u + 10y + x$ with $nx \approx v$ and $nv - x \equiv 0$ mod 10.

Only for $n = 4, 9$ do we get a possible solution: $n = 4$: $4v - x \equiv 0$ mod 10 and $4x \approx v$; $x = 2$, $v = 8$ satisfies, so: $13y + 1 = 2u$ yielding $y = 1$, $u = 7$, cost = \$21.78. $n = 9$: $9v - x \equiv 0$ mod 10, $9x \approx v$, $x = 1$, $v = 9$ satisfies and $90y + 8 = u$ yielding $y = 0$, $u = 8$, so cost = \$10.89. In view of the restriction against 4 purchases, only the latter solution is acceptable.

A Diophantine Equation

Rearranging, $y^2 - x^2y - 4x^2 = 0$. Solving as a quadratic in y,

$$y = \frac{x^2 \pm \sqrt{x^4 + 16x^2}}{2} = \frac{x}{2}(x \pm \sqrt{x^2 + 16})$$

Since y is an integer, the radicand must be a perfect square, hence $x = 0$ or ± 3, giving the solutions $(x, y) = (0, 0)$, $(\pm 3, -3)$, or $(\pm 3, 12)$.

Finding the Coefficients

Let
$$x^2 + Ax + B = (x + r)(x + s)$$
and
$$x^2 + Bx + A = (x + t)(x + u)$$

Then $A = r + s = tu$ and $B = rs = t + u$. Since $A < B$, $tu < t + u$. If both t and $u \geq 2$, then $tu \geq 2 \max (t, u) \geq t + u$, a contradiction.

Hence either t or $u = 1$. Thus $rs = r + s + 1$. Neither r nor s can equal 1 and if both exceed 2, then $rs \geq 3$ max $(r, s) > r + s + 1$, another contradiction. Hence one of them, say r, is 2 and $2s = s + 3$ or $s = 3$. We then have $A = 5$; $B = 6$ as the only solution.

Another Diophantine Equation

We use the fact that the sum or difference of two multiples of a number is a multiple of that number. We note that $x + 3$ is a factor of $x^2 + 2$. Since it is also a factor of $x^2 + 3x$, it is a factor of $x^2 + 3x - (x^2 + 2) = 3x - 2$. Since it is also a factor of $3x + 9$, it is a factor of $(3x + 9) - (3x - 2) = 11$. Now, since x is positive, $x + 3 = 11$ and $x = 8$; $y = 6$.

4

THE DATA SEEKERS

Problems in Logic and Deduction

A Digital Arrangement (SOLUTION, Page 94)
Finding the Resistors (SOLUTION, Page 98)
All the Digits (SOLUTION, Page 100)
No Change · Subtract-a-square · An Alphametic · Husbands and Wives · No Repetitions Allowed · Ups and Downs · Chess Mates · Poetry in Motion · Seven Towns · The Golf Tournament · A Family Reunion · The Cock Robin Murder · Choosing Socks · A Penny Raise · A Time and Motion Problem · The Eternal Octagon · The Cheating Grocer · Martian Algebra · The Maximal Product · Loose Wires · A Professional Gathering · A Committee Problem · A Weight Problem · An Unusual Year · Creating Palindromes · The Seven-digit Problem · The Pastoral Lovers · Getting Back in Step · The Photographer · The Wizard

SOLUTIONS: Page 129

A DIGITAL ARRANGEMENT

Without using any symbols, arrange the digits 1, 3, 5, 7, 9 to equal the digits 2, 4, 6, 8.

Solution

$$(35^1)_{(7_9)} = (42)_{(6_8)}$$

The subscripts, indicating the notational base in which the number is represented, baffled several readers. In the above solution, $7_9 = 7$ and $6_8 = 6$. Therefore, $(35^1)_{(7_9)} = 35_7 = 26$ and $(42)_{(6_8)} = 42_6 = 26$.

Among the responses to this problem were some of the more original we have received. Certainly the most concise came from an M.D. in Wisconsin who wrote simply "Help! Help!" From some gentlemen at Cal-Tech's Jet Propulsion Laboratory we received this appeal:

> A group of us have been following your Problematical Recreations with intrepid enthusiasm until we reached last week's problem. This problem baffled our brains completely. We therefore eagerly sought its solution in the next issue and promptly launched into a state of despair.... We would appreciate any enlightenment which you can give us on this answer so that we can then proceed to the next problem with renewed, uncompromised vigor.

From an assistant project manager at Pratt & Whitney Aircraft in Connecticut:

> I read and enjoy your weekly problems but this one throws me ... what does the notation mean? The 1, I guess, is an exponent. The x_y isn't recognized by any of our mathematicians. We tried binomial coefficients but they don't seem to fit. So, gentlemen, please wotinell does it mean?

From Fairfax, Virginia:

> Please explain the use of the subscripts. I have discussed this answer with several mathematicians and they have not been able to satisfy me.

Then from Poughkeepsie, New York, came an alternate solution which is not only more elegant than ours but adheres

94

more strictly to the restrictions in the statement of the problem in that no parentheses were used:

$$39\frac{1}{57} = 264_8$$

We congratulated the sender, Russell Adem, and wished we had thought of it first.

FINDING THE RESISTORS

An engineer ordered 9 boxes of 100-ohm resistors and 1 box of 110-ohm resistors. When they arrived there were 10 resistors in each of the 10 boxes, but both the boxes and the resistors were unmarked. How many resistance measurements did he have to make to locate the box of 110-ohm resistors?

Solution

One. He took 1 resistor from the 1st box, 2 from the 2nd, etc., making 55 in all. He then measured the resistance of all 55 connected in series. Deducting 5500 from the total and dividing the difference by 10 gives the number of the box containing the 110-ohm resistors.

Our own variation of this well-known programming problem brought some resistance from a very practical gentleman in Chicago who asked, "Is mathematics the servant or the master of the engineer?" and offered the following advice:

> . . . I bet I could find his 110-ohm box in half the time. Seems to me that a random check of resistors in the 10 boxes would bring the quickest results. In the first place, if I remember right, these resistors are accurate only within plus or minus 5 or 10 per cent. If this is still true, it throws his calculations out of whack. Chances are fairly good that many 100-ohm resistors would measure the same as many 110-ohm resistors. Certainly 55 of them mixed would hardly total their collective ratings to any degree of accuracy. In the second place, connecting 55 of them in series is a bigger job than checking random boxes. Finally, whatinhell is the poor guy going to do now that he knows the number of the box? He's got more than half those damn things out of their boxes connected in series. If he forgot to mark carefully which resistors came from which box (and he sounds like the kind of dolt who would forget), he's still faced with a big measuring job.

We commended the writer for his practicality but hastened to add that we were offering an exercise in logical thinking, not a practical guide for EE's. Besides, the problem demands determining with certainty the required box, whereas the suggested random check method (practical as it may be) can at best determine this box with a high probability. Concerning the accuracy of resistors in general, we plead ignorance. As is true in many mathematical problems, ideal conditions must be blithely assumed.

ALL THE DIGITS

Arrange the digits 0 through 9 in fractional form so that:

$$\frac{xx,xxx}{xx,xxx} = 9.$$

Solution

$$\frac{97,524}{10,836} = 9 \quad \text{or} \quad \frac{75,249}{08,361} = 9, \quad \text{or} \quad \frac{57,429}{06,381} = 9$$

Our three answers had been in print but a few days when we received the following additional solution:

$$\frac{58,239}{06,471} = 9$$

We had just recovered from our embarrassment when several readers brought yet another valid answer to our attention:

$$\frac{95,742}{10,638} = 9$$

Now there were five, we had to agree, feeling slightly uneasy. At this point we began to expect almost anything so we were not too surprised to find a sixth valid answer in the day's mail:

$$\frac{95,823}{10,647} = 9$$

It is interesting to note that we received every combination of one or two additional solutions and five thorough gentlemen sent all three. We feel it is only fitting to mention them by name: Fred Mickey, Dick Rentfrow, and Ed Katenkamp of Los Angeles, and J. Bolze and F. Nassauer of New York.

For the benefit of those ambitious readers who might attempt to find a seventh solution, we would like to add that the six have been shown to be exhaustive.

NO CHANGE

What is the largest amount of money you can have in coins and still not be able to give change for a dollar?

For solutions to problems in this section, turn to page 129.

SUBTRACT-A-SQUARE

In the game "subtract-a-square," a positive integer is written down and two players alternately subtract squares from it until a player is able to leave zero, in which case he is the winner. What square should the first player subtract if the original number is 29?

AN ALPHAMETIC

Show that two WRONG's *can* make a RIGHT, even with the additional restriction that 0 = zero.

HUSBANDS AND WIVES

In a certain community there are 1000 married couples. Two-thirds of the husbands who are taller than their wives are also heavier and three-quarters of the husbands who are heavier than their wives are also taller. If there are 120 wives who are taller and heavier than their husbands, how many husbands are taller and heavier than their wives?

A man has red, gray, and black flagstones for making a walk. He wants no two consecutive stones to be the same color, no consecutive pair of stones to have the same two colors in the same order, no repetition of three consecutive colors, etc. He starts out laying first a red stone, then a gray, and continues until he finishes laying the seventh stone. He then finds himself stymied and unable to use any stone for the eighth without repetition of some color pattern. What were the colors of the first seven stones?

UPS AND DOWNS

Smith and Jones started working for different firms at the same salary. Last year Smith had a raise of 10 per cent and Jones had a drop in pay of 10 per cent. This year Smith had the 10 per cent drop and Jones the 10 per cent raise. Who is making more now?

CHESS MATES

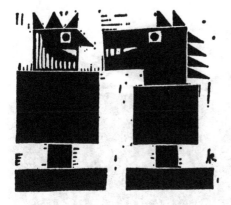

A modernistic chess set has pieces in various geometrical shapes. In particular, both the KING and the KNIGHT are squares of integers. What numbers could these represent if each letter is replaced by a different digit?

POETRY IN MOTION

Noel Wentworth-Longmore, the famous Oxford rower, was rowing upstream on Sunday when his favorite rowing cap fell in the water. So absorbed was he in one of Housman's poems that it was 10 minutes before he discovered his cap missing. He turned around and recovered the cap 1 mile downstream from where he initially lost it. Assuming constant speed and no allowance for turnaround, how fast was the river flowing?

SEVEN TOWNS

Between Kroflite and Beeline are five other towns. The seven towns are an integral number of miles from each other along a straight road. The towns are so spaced that if one knows the number of miles a person has traveled between any two towns he can determine the particular towns uniquely. What is the minimum distance between Kroflite and Beeline to make this possible?

THE GOLF TOURNAMENT

An elimination golf tournament is held for 300 players. How many golf matches will transpire?

A FAMILY REUNION

The family of a quality control engineer consisted of 1 grandmother, 1 grandfather, 2 fathers, 2 mothers, 4 children, 3 grandchildren, 1 brother, 2 sisters, 2 sons, 2 daughters, 1 father-in-law, 1 mother-in-law, and 1 daughter-in-law. What is the smallest possible number of persons in his family?

THE COCK ROBIN MURDER

Five suspects were rounded up in connection with the famous "Cock Robin Murder." Their statements were as follows: *A:* "*C* and *D* are lying." *B:* "*A* and *E* are lying." *C:* "*B* and *D* are lying." *D:* "*C* and *E* are lying." *E:* "*A* and *B* are lying." Who is lying?

CHOOSING SOCKS

A drawer contains an odd number of plain brown socks and an even number of plain black socks. What is the least number of brown and black socks such that the probability of obtaining two brown socks is 1/2 when two socks are chosen at random from the complete collection?

A PENNY RAISE

A canoe is floating in a swimming pool. Which will raise the level of the water in the pool higher, dropping a penny into the pool or into the canoe? Or does it make any difference?

A TIME AND MOTION PROBLEM

An engineer must test three space suits in two test chambers. Each suit must be tested for 1 hour at each of two low pressures. He takes 10 minutes to load a suit in a chamber, set the pressure, and start the test; 4 minutes to change the pressure; and 10 minutes to unload a suit from a chamber. What is the minimum time to complete the tests?

THE ETERNAL OCTAGON

Four boys, Alan, Brian, Charles, and Donald, and four
girls, Eve, Fay, Gwen, and Helen are each in love with
one of the others and, sad to say, in no case is their love
requited. Alan loves the girl who loves the man who loves
Eve. Fay is loved by the man who is loved by the girl
loved by Brian. Charles loves the girl who loves Donald.
If Brian is not loved by Gwen, and the boy who is loved
by Helen does not love Gwen, who loves Alan?

THE CHEATING GROCER

Six grocers in a town each sell a different brand of tea in 4-ounce packets at 25 cents per packet. One of the grocers gives short weight, each packet of his brand weighing only $3\frac{3}{4}$ ounces. If I can use a scale for only one weighing, what is the minimum amount I must spend to be sure of finding the grocer who gives short weight?

MARTIAN ALGEBRA

The first expedition to Mars found only the ruins of a
civilization. The explorers were able to translate a
Martian equation as follows: $5x^2 - 50x + 125 = 0$:
$x = 5$ or 8. This was strange mathematics. The value
$x = 5$ seemed legitimate enough but $x = 8$ required
some explanation. If the Martian number system
developed in a manner similar to ours, how many fingers
would you say the Martians had?

THE MAXIMAL PRODUCT

What is the largest number which can be obtained as
the product of positive integers which add up to 100?

LOOSE WIRES

A man has an odd number of wires running from his basement to his roof. He has available some numbered tags and a meter which detects open and short circuits. In order to label the corresponding ends of the wires in the basement to those on the roof, what is the least number of round trips necessary?

A PROFESSIONAL GATHERING

A bunch of the chaps from Bristol dropped into a nearby pub. There were four parties: 25 physicists, 20 engineers, 18 productivity managers, and 12 comptrollers. Altogether they spent 6 pounds sterling, 13 shillings.* It was found that 5 physicists spent as much as 4 engineers; that 12 engineers spent as much as 9 productivity managers; and that 6 productivity managers spent as much as 8 comptrollers. How much did each of the four parties spend?

*At the rate of exchange of 1 pound sterling = 20 shillings = $2.80.

A COMMITTEE PROBLEM

Ann, Barbara, Carol, and Dorothy are members of the Bobby-Sox Club. Every pair of members is together on one and only one committee. Each committee has exactly 3 members. What is the smallest possible total membership, and how many committees are there?

A WEIGHT PROBLEM

There are five weights, no two weighing the same. With a beam balance, arrange the weights in order from heaviest to lightest in seven weighings.

AN UNUSUAL YEAR

The year 1961 had the rare property of reading the same upside-down. Of how many years (A.D.) has this been true, and how many more will elapse before another?

CREATING PALINDROMES

If a number is added to its reversal and the process repeated with the result, a number will eventually be obtained which reads the same backward and forward, i.e., is "palindromic." For a certain two-digit number this process must be repeated more than ten times to arrive at a palindromic number. What is this number?

THE SEVEN-DIGIT PROBLEM

There are 120 seven-digit numbers which can be formed by starting from any number in the above diagram and proceeding to any neighboring number, using each number once and only once. Of these, how many are divisible by 11?

THE PASTORAL LOVERS

Strephon and Phyllis decide to test their love with a daisy. They agree to pluck petals alternately, taking either one petal or two adjacent petals. There are 13 petals altogether. He picks one saying "She loves me." She picks two adjacent petals, leaving two groups of 8 and 2, saying, "He loves me not." How should Strephon continue?

GETTING BACK IN STEP

A husband and wife start walking together, both stepping out on the right foot. The wife takes $k + 1$ steps for every k steps her husband takes and in walking a mile she takes 531 steps more than her husband. How many times in walking that distance did they simultaneously step out on the left foot?

THE PHOTOGRAPHER

For a group picture a photographer wishes to arrange 10 people (all of different heights) in two rows of five each. Each person in the back row must be taller than the person in front of him. Also, the various heights are to increase in each row from left to right. In how many ways can this be done?

THE WIZARD

Not having colored ink with which to make red and black dots, the Wizard wrote numbers on the foreheads of his three apprentices instead and announced that each had been given a prime number (not necessarily distinct) and that the three numbers formed the sides of a triangle with prime perimeter. The first to deduce his own number was to be the Wizard's successor. Apprentice A noted that B had a 5 and C had a 7. After a long period of silence he announced his number. What was it?

Solutions

No Change

$1.19. Three quarters, four dimes, and four pennies.

Subtract-a-square

To ensure a win the first player must subtract 9. He then counters his opponents' plays of 1, 4, 9, or 16 with 9, 16, 9, or 4 respectively.

An Alphametic

$$
\begin{array}{ccc}
37081 & & 37091 \\
\underline{37081} & \text{and} & \underline{37091} \\
74162 & & 74182
\end{array}
$$

Husbands and Wives

If we let X = number of husbands taller than their wives,

Y = number of husbands heavier than their wives,

and Z = number of husbands taller and heavier than their wives, we have $Z = \frac{2}{3}X$, $Z = \frac{3}{4}Y$ and $1000 - X - Y + Z = 120$.

From the first two equations, $8X = 9Y$, which yields, in conjunction with the third equation, $X = 720$, $Y = 640$, so that the number of husbands taller and heavier than their wives is 480.

No Repetitions Allowed

R G R B R G R. He now cannot use R for then there would be two consecutive R's. He cannot use G because there would be two consecutive R G's. He cannot use B for then there would be two consecutive R G R B's. Any other pattern for the first seven stones would have allowed a choice for the eighth not involving a repetition.

Ups and Downs

Let x = the original salary. After Smith's raise his salary was $\frac{11x}{10}$.

After his drop it was $\frac{9}{10} \cdot \frac{11x}{10}$. After Jones' drop his salary was $\frac{9x}{10}$ and after his raise it was $\frac{11}{10} \cdot \frac{9x}{10}$. Thus both are presently making $\frac{99}{100}$ times the original salary.

Chess Mates

The minimum value for $\sqrt{\text{KING}}$ is 32, the maximum 99. Between these limits there are 36 values the squares of which have no repeated digits. Rearranging these as the first four digits of KNIGHT, we find 327184 as the only such six-digit square with no repeated digits. Therefore KING = 3721, KNIGHT = 327184.

Poetry in Motion

Since the rower pursues his cap at the same rate he rowed away from it, to wit, his rowing speed in still water, it takes him another 10 minutes to recover it. Since the cap floats 1 mile in 20 minutes, the river flows at the rate of 3 mph.

Seven Towns

The distance from Kroflite to Beeline must be at least 25 miles. The towns could then be located at distances 0, 1, 4, 10, 18, 23, and 25 miles from Kroflite. There are 21 distances between towns and these are all distinct. Any shorter distance would mean at least one duplication.

The Golf Tournament

Since there is only one winner, there must be 299 losers, requiring 299 matches.

A Family Reunion

Seven. Two little girls and a boy, their father and mother, and their father's father and mother.

The Cock Robin Murder

If we assume B is telling the truth, then by following the implication of his statement we find that D is also telling the truth. If we assume B is lying, we find that C and E are telling the truth. In either event, however, A is lying. Thus A is the only suspect we know with certainty to be lying.

Choosing Socks

15 brown and 6 black socks.

130

A Penny Raise

It does make a difference. A submerged body displaces its volume; a floating body displaces its weight. Since a penny is denser than water, dropping it into the canoe will raise the water level higher.

A Time and Motion Problem

3 hours 54 minutes. Load suit A into 1st chamber, load suit B into 2nd, wait 50 minutes, change pressure in 1st, wait 6 minutes, unload suit B from 2nd, load suit C into 2nd, wait 34 minutes, unload A from 1st, load B into 1st, wait 6 minutes, change pressure in 2nd, wait 50 minutes, unload B from 1st, unload C from 2nd.

The Eternal Octagon

Let $x \rightarrow y$ denote "x loves y" and $x \nrightarrow y$ denote "x does not love y." We have then:

$$(1) \quad A \rightarrow ? \rightarrow ? \rightarrow E \rightarrow ?$$
$$(2) \quad B \rightarrow ? \rightarrow ? \rightarrow F \rightarrow ?$$
$$(3) \quad C \rightarrow ? \rightarrow D$$
$$(4) \quad G \nrightarrow B \quad \text{and}$$
$$(5) \quad H \rightarrow ? \nrightarrow G$$

The condition that no love is requited implies that either there is a closed circle consisting of all eight individuals, or else there are two closed circles consisting of four individuals each. Assume the latter is true. Then since C and D are in the same circle [from (3)] A and B must be in the same circle. We get $A \rightarrow F \rightarrow B \rightarrow E \rightarrow A$. Now since G and H are together in the other circle, (5) is violated. Hence all eight individuals are in the same circle. Now the lover of E is not A [condition (1)], nor D [condition (3)] nor C, which eventually contradicts (5). Hence B loves E. This enables us to complete the circle: $A \rightarrow H \rightarrow B \rightarrow E \rightarrow C \rightarrow F \rightarrow D \rightarrow G \rightarrow A$. We conclude that Gwen loves Alan.

The Cheating Grocer

$3.75. Buy no packets from the first grocer, one from the second, two from the third, three from the fourth, four from the fifth, and five from the sixth. If the total weight amounts to 60 ounces, then the grocer from whom I did not purchase a packet gives short weight. If the total weight is $59\frac{3}{4}$ ounces, then the grocer from whom I purchased one packet gives short weight, etc.

Martian Algebra

We shall assume that the base of the number system is equal to the number of fingers. If b is the base then we can write the equation as follows:

$$5x^2 - 5bx + (b^2 + 2b + 5) = 0$$

Thus $b = 5 + 8 = 13$ and the Martians had 13 fingers.

The Maximal Product

Clearly 1 would not appear as a factor, and any 4 could be replaced by two 2's, without decreasing the product. And if one of the factors were greater than 4, replacing it by 2 and $n - 2$ would yield a larger product. Thus the factors are all 2's and 3's. Moreover, not more than two 2's are used, since the replacement of three 2's by two 3's would increase the product. The largest number possible is therefore $3^{32} \cdot 2^2$.

Loose Wires

One round trip. At the basement, short-circuit pairs of wires, leaving one odd wire. At the roof the odd wire is found and labeled 1. It is then shorted to any wire which is then labeled 2. The wire to which 2 was shorted at the basement is found and labeled 3. Wire 3 is shorted to any remaining wire which is labeled 4. Wire 5 is found as the wire to which 4 was shorted in the basement, etc. Then, at the basement, the original short circuits are opened. Wire 1 is the odd wire. Wire 2 is found by finding the wire shorted to 1. Wire 3 had been originally shorted to 2 in the basement. Wire 4 is found by finding the wire shorted to 3, etc.

A Professional Gathering

Using productivity managers as our basic unit, the 12 comptrollers spend as much as 9 productivity managers, the 20 engineers as much as 15 productivity managers, and the 25 physicists as much as 20 engineers, hence as much as 15 productivity managers. Thus the equivalent of 57 productivity managers spent a total of 133 shillings. Of these, the physicists spent $\frac{15}{57} \cdot 133$ shillings = 1 pound, 15 shillings. The engineers spent a like sum. The productivity managers spent $\frac{18}{57} \cdot 133$ shillings = 2 pounds, 2 shillings and the controllers $\frac{9}{57} \cdot 133$ shillings = 1 pound, 1 shilling.

A Committee Problem

It is quickly seen that the club must have more than the 4 members named. The easiest method is to start building up the committees,

132

and the solution is found to be 7 members and 7 committees, which can be designated 123, 145, 167, 246, 257, 347, and 356.

A Weight Problem

Weigh A against B, C against D, then match the two heavier ones. Without loss of generality assume A heavier than B and C, with C heavier than D. This can be accomplished in 3 weighings. Now weigh E against C, and suppose E heavier than C, in which case weigh E against A (5th weighing). If A is heavier than E, then match B against C, followed by a match between B and E or B and D depending on whether B is heavier or lighter than C. If E is heavier than A, match B against C and against D if necessary, which completes the ordering in 7 or fewer weighings. The case of E lighter than C can be handled in a like manner.

Whether the ordering can be accomplished in fewer than seven weighings is doubtful, but still an open question.

An Unusual Year

It has been true 24 times since and including the year zero. However, over 4000 years more must elapse until the next occurrence which takes place in 6009.

Creating Palindromes

Any number with two different digits, having the sum of its digits less than 10 will produce a palindromic number in one operation. We need consider therefore only numbers the sum of whose digits is 10 or greater. If the sum of the digits is 10, 11, 12, 13, 14, 15, or 16, the number of operations required will be 2, 1, 2, 2, 3, 4, 6 respectively. Only when the sum of the digits is 17 are more than 10 operations required. But the only number having 17 as the sum of its digits is 89 (or its reversal, 98).

The Seven-digit Problem

For a number to be divisible by 11, the sum of the odd-placed digits must be congruent (mod 11) to the sum of the even-placed digits. Since the sum of all seven digits is 28, we seek partitions of 14 into three and four parts, using the numbers 1 to 7 once and only once. There are exactly 10 such numbers subject to the given conditions: 1237456, 1237654, 1675432, 2345761, 3217456, 3217654, 4567123, 4567321, 6547123, and 6547321.

The Pastoral Lovers

Strephon can win by picking one of the end petals from the group of 8 making two groups of 7 and 2. Then if Phyllis leaves 7, he can make this 3, 3, and if she leaves 2, 2, 4, he can make this 2, 2, 1, 1, in both cases winning since he can duplicate her later moves. If she leaves anything else, he can convert it into 1, 2, 3, and win. Strephon can also win by leaving groups of 8 and 1, groups of 4, 3, and 2, or groups of 6, 1, 2, as pointed out by an astute contributor. The reader will enjoy determining the moves and countermoves for these three alternate solutions.

Getting Back in Step

They will first be in step again after the wife has taken $k + 1$ and the husband k steps. Since one of these numbers is odd, the other even, one will step out with the left foot, the other with the right. They will next be in step after the wife has taken $2k + 2$, the husband $2k$ steps. Since both numbers are even, they will both step out with their right feet again. This pattern obviously repeats. Hence they never step out on the left foot simultaneously.

The Photographer

Number the people from 1 to 10 in order of increasing height. Then the first man in the front row must be 1 and the last man in the back row must be 10. The last man in the front row must be 5, 6, 7, 8, or 9. There are respectively 1, 4, 9, 14 and 14 ways of filling in the other numbers in the front row. Once the front row has been selected, the numbers in the back row are completely determined. There are thus $1 + 4 + 9 + 14 + 14 = 42$ ways of arranging the people.

The Wizard

A reasoned as follows: "My number is between 3 and 11 in order to satisfy the triangle inequality. Since it is prime, it is either 3, 5, 7, or 11. 3 can be rejected since it would make the perimeter composite. 5 can be rejected because if I had 5, C would reason this way: 'I see two 5's, so I have either a 3 or a 7. But if I had a 3, then A and B could each deduce that they had 5's, since if either had a 3, the other would know that he had to have a 5. Since neither has announced his number I don't have a 3. Hence I have a 7.' Since C has not so announced, I don't have a 5. If I had a 7, B would reason thus: 'I see two 7's and, therefore, I must have either a 3 or a 5. But if I had a 3, A and C would know immediately that they each had 7. Since neither has so announced, I must have a 5.' Since B has not so announced, I can't have a 7. Hence my number is 11."

5

MINDING YOUR P'S AND Q'S

Probability Posers

The Alpenstock (SOLUTION, Page 138)

Odd Numbers · Points on a Sphere · Increasing License Numbers · A Round Trip · A Game of Chess · Odd Man Out · A Full Deck · The Spider and the Fly · Birthday Odds · Throwing a Die · One of Each · The Lost Coin · The Escaping Convict

SOLUTIONS: Page 149

THE ALPENSTOCK

Johann Jungfrau, the famous mountain climber, was traveling through the Trondheim timber country one day. Quite by accident he dropped his trusty alpenstock, an unusually straight stick, near the buzz saws where, in two shakes of a yak's tail, it was neatly cut into three pieces. What is the probability that these three pieces can be placed together to form a triangle?

Solution

A triangle can be formed if and only if: (1) the breaks are on opposite sides of the midpoint of the alpenstock (probability $\frac{1}{2}$) and (2) the breaks are less than $\frac{1}{2}$ an alpenstock's distance apart (probability $\frac{1}{2}$). The desired probability is therefore $\frac{1}{4}$.

Jungfrau was revisited by his friends, Ernest Cohen and Meyer B. Shulman of the G.E. Missile and Space Division in Philadelphia, who submitted a variation which at the surface may seem to be the same as our problem, but is actually different. We offer it for your consideration:

Johann Jungfrau was again traveling through the Trondheim Timber Country with a new trusty alpenstock. As luck would have it, he again dropped his alpenstock into a buzz saw, which cut it into two pieces. He was so angered by his repeated accident that he threw one of the pieces at random into the buzz saw, which neatly cut it into two. What is the probability that the three pieces can be placed together to form a triangle?

ANSWER

The three pieces will form a triangle if no one part is greater than one-half the original stick. This condition can be satisfied if the second cut is performed on the larger piece (probability $\frac{1}{2}$) such that one of the pieces is not greater than one-half the stick.

Using a normalized value of 1 for the length of the random piece, with x as the length of the smaller cut piece and $1 - x$ the length of the larger piece, the probability for a triangle to occur will be equal to

$$\frac{1}{2} \int_0^{\frac{1}{2}} \frac{2x}{(1 - x)} \, dx$$

where the factor 2 is inserted in the integral since this condition can occur at either side of the center of the normalized piece.
Now:

$$\int_a^b \frac{x \, dx}{1 - x} = \left[(1 - x) - \ln |1 - x| \right]_a^b$$

Thus, the answer is $\ln 2\frac{1}{2} = .1931$.

ODD NUMBERS

The numbers one through seven are drawn from a hat without replacement. What is the probability that all the odd numbers will be chosen first?

For solutions to problems in this section, turn to page 149.

POINTS ON A SPHERE

Three marksmen simultaneously shoot at and hit a rapidly spinning spherical target. What is the probability that the three points of impact are on the same hemisphere?

INCREASING LICENSE NUMBERS

On a certain day, our parking lot contains 999 cars, no two of which have the same three-digit license number. After 5:00 P.M., what is the probability that the license numbers of the first four cars to leave the parking lot are in increasing order of magnitude?

A ROUND TRIP

The planet Octerra is divided into eight countries, each occupying an octant. (Thus each country borders three others.) In how many ways can a traveler visit each of the other countries once and only once, returning to his home country only at the end of his trip?

A GAME OF CHESS

A castle and a bishop are placed at random on different squares of a chessboard. What is the probability that one piece threatens the other?

ODD MAN OUT

Smith and Jones, both 50 per cent marksmen, decide to fight a duel in which they exchange alternate shots until one is hit. What are the odds in favor of the man who shoots first?

A FULL DECK

Assuming that each pack of cigarettes from a certain manufacturer contains, as a premium, one of a set of 52 playing cards and that these cards are distributed among the packs at random (the number of packs available being infinite), what is the expected number of packs that must be purchased in order to obtain a complete set of cards?

THE SPIDER AND THE FLY

A spider eats three flies a day. Until he fills his quota, he has an even chance of catching any fly that attempts to pass. A fly is about to make the attempt. What are the chances of survival, given that five flies have already made the attempt today?

BIRTHDAY ODDS

If 23 people are in a room, the chances are better than even that at least two were born on the same day of the year. How many must be present in order to provide at least an even chance that two or more were born on the same day of the week?

THROWING A DIE

When a single die is thrown, what is the probability that the expected (in the statistical sense) number will come up?

ONE OF EACH

What is the expected number of children a couple must have in order to have both a boy and a girl?

THE LOST COIN

After having his dollar changed into six coins, a boy had the misfortune to drop one of them down a grating. What is the chance that the coin was a dime?

THE ESCAPING CONVICT

Smith's chance of hitting his target on a given shot is twice that of his fellow prison guard, Jones. One dark night they each had time to fire one shot at an escaping prisoner. Given that the prisoner had an even chance of avoiding injury, what kind of a marksman was Jones?

Solutions

Odd Numbers

The desired probability is $\frac{3}{7} \cdot \frac{2}{8} \cdot \frac{1}{5} = \frac{1}{35}$.

Points on a Sphere

The probability is one, since any three points on the surface of a sphere are always located on some hemisphere.

Increasing License Numbers

There are 4! or 24 possible permutations of 4 cars. Only one of these is in increasing order of license magnitude. Thus there is one chance in 24. The number of cars in the lot (999) is irrelevant.

A Round Trip

Starting from his home country the traveler has three possible choices. After visiting the second country he has two choices, and after visiting the third country, two more choices. But after the first four countries have been visited, the second half of his trip is uniquely determined. There are, therefore, $3 \times 2 \times 2 = 12$ possible ways of making the trip.

A Game of Chess

If the castle is on one of the 4 center squares, it threatens 14 squares and is threatened diagonally by 13 squares for a total of 27. This total is 25 for the squares bordering the center, 23 for the squares bordering these, and 21 for the outer border. The probability is therefore

$$\frac{4}{64} \cdot \frac{27}{63} + \frac{12}{64} \cdot \frac{25}{63} + \frac{20}{64} \cdot \frac{23}{63} + \frac{28}{64} \cdot \frac{21}{63} = \frac{13}{36}$$

Odd Man Out

The first man's odds are

$$\frac{1}{2} + \frac{1}{2} \cdot \frac{1}{2} \cdot \frac{1}{2} + \frac{1}{2} \cdot \frac{1}{2} \cdot \frac{1}{2} \cdot \frac{1}{2} \cdot \frac{1}{2} + \cdots$$

$$= \frac{1}{2} \left(1 + \frac{1}{4} + \frac{1}{4^2} + \frac{1}{4^3} + \cdots \right) = \frac{1}{2} \left(\frac{4}{3} \right) = \frac{2}{3}$$

149

A Full Deck

On the average, the entire set of 52 cards will be obtained in

$$52\left(\frac{1}{52} + \frac{1}{51} + \cdots + \frac{1}{3} + \frac{1}{2} + 1\right)$$

trials. Hence the average number of trials will be

$$52 \sum_{n=1}^{52} \frac{1}{n} \quad \text{or} \quad 235.976$$

and since we must buy an integral number of cigarette packs, the number will be 236.

The Spider and the Fly

The chance that the first three flies were captured is $(\frac{1}{2})^3 = \frac{1}{8}$. The chance that the fourth fly completed the quota is $\frac{\binom{3}{2}}{2^4} = \frac{3}{16}$. The chance that the fifth fly completed the quota is $\frac{\binom{4}{2}}{2^5} = \frac{6}{32}$. Thus the probability that the spider has completed his daily fare is $\frac{1}{8} + \frac{3}{16} + \frac{6}{32} = \frac{1}{2}$. Hence the fly has a $\frac{1}{2}$ chance of being attacked and a $\frac{1}{4}$ chance of being captured. His survival probability is therefore $\frac{3}{4}$.

Birthday Odds

If three people are present, the chance that at least two were born on the same day of the week is only $\frac{19}{49}$ or about 0.39. The answer is four people, in which the probability increases to $\frac{223}{343} \cong 0.65$.

Throwing a Die

Zero. The expected number is

$$\frac{1 + 2 + 3 + 4 + 5 + 6}{6} = 3\frac{1}{2}$$

One of Each

Three. The parents first have children of both sexes after the birth of the nth child provided the first $n - 1$ children are of the same sex and the nth child is of the opposite sex. This can happen in two ways

150

out of a possible 2^n. The probability is therefore $\dfrac{1}{2^{n-1}}$ and the expected number of children is:

$$\sum_{n=2}^{\infty} \frac{n}{2^{n-1}} = 3$$

The Lost Coin

There are three ways of changing a dollar into six coins: half, quarter, dime, 3 nickels; half, 5 dimes; and 3 quarters, 2 dimes, nickel. Information lacking, we must assign probability $\frac{1}{3}$ to each of these. The probability that the coin was a dime is therefore $\frac{1}{3}\cdot\frac{1}{6} + \frac{1}{3}\cdot\frac{5}{6} + \frac{1}{3}\cdot\frac{2}{6} = \frac{4}{9}$. In similar fashion the respective probabilities for half, quarter, and nickel are found to be $\frac{1}{9}$, $\frac{2}{9}$, and $\frac{2}{9}$.

The Escaping Convict

A poor one. Letting Jones' probability be p, in which case Smith's is $2p$, the chance that both miss the prisoner is $(1 - p)(1 - 2p) = \frac{1}{2}$, leading to the equation $4p^2 - 6p + 1 = 0$ with roots $p = \dfrac{3 \pm \sqrt{5}}{4}$. Since $p < 1$, the positive choice is rejected, and

$$p = \frac{3 - \sqrt{5}}{4} \doteq .19$$

6

NOW YOU SEE IT......

Insight Puzzles

The Campers (SOLUTION, Page 156)

A Close Decision (SOLUTION, Page 158)

The Missing Number (SOLUTION, Page 160)

A Familiar Sequence · *A Mathematical Quickie* · *A Dearth of Primes* · *Cheaper by the Thousand* · *Toe-tac-tic* · *Exchanging Hands* · *Counting Primes* · *Efficient Weighing* · *An Intercity Meeting* · *A Flag Problem* · *A Unique Digital Property* · *Number Classes* · *A Peculiar Progression* · *The Clever Grandfather* · *The Gold Chain* · *The Pigeonholes* · *Computing Area without Integration*

SOLUTIONS: Page 177

THE CAMPERS

Two campers, Smith and Jones, have pitched their tents not far from a river. Smelling smoke, Smith looks out of his tent and sees Jones' tent on fire. Smith grabs a bucket and is about to make a dash for the river to fill the bucket followed by a quick return to Jones' tent. The chances of quenching the fire are good if he wastes no time. Can you advise Smith what point of the river he should head for?

Solution

Smith should imagine Jones' tent reflected through the line of the river to the opposite side and should head directly for that point, stopping, of course, when he gets to the river. A simple indirect proof will show that this minimizes the total distance.

If we would get out and go camping once in a while instead of keeping our nose to the slide rule, we would know that a full bucket of water slows one down a bit, a factor we overlooked. Our readers were obviously all experienced bucket carriers and fire fighters.

A Navy man summed up the situation concisely: "Since he will be returning from river with bucket full of water, return trip should be somewhat shorter to compensate for reduced speed due to increased weight."

From Sioux Falls, South Dakota, came the following more expanded explanation:

> Let's be practical about this situation! Nothing is said in the problem about distance; time is what we want to save. Smith may or may not be very strong, and he may or may not have a pretty good-sized bucket. At any rate the smart thing for him to do is to run to a point on the river that is the shortest distance from Jones' tent with his empty bucket and then haul the heavy bucket of water this shortest distance. The poor little guy may not ever get there if he has to haul that heavy bucket any further.

There is little we can add to the above letters.

A CLOSE DECISION

Which is larger, the tenth root of 10 or the cube root of 2?

Solution

Use of log table or slide rule involves unnecessary inaccuracy and time consumption. Simply raise both numbers to the 30th power, giving 1000 and 1024 respectively. Hence the cube root of 2 is larger.

In spite of the published solution, we received a minor deluge of mail pointing out that an alternate solution requiring simply the fact that the common logarithm of 2 is greater than .3 exists, to wit, that $\log 3\sqrt{2} = .1^{+}$, while $\log {}^{10}\sqrt{10} = \frac{1}{10} = .1$ exactly. Hence ${}^{3}\sqrt{2} > {}^{10}\sqrt{10}$.

We don't deny the merit of this proof but still assert the essential point that the simplest approach is frequently the most direct. Several of the problems in this chapter are excellent examples of this principle.

There is a special beauty in a solution which is at once clear and concise, though far from obvious. Mathematicians refer to such solutions as "elegant." We hope you find several elegant solutions of your own, not forgetting, however, that even an inelegant solution is better than no solution at all.

THE MISSING NUMBER

Supply the missing number in the following sequence:
10, 11, 12, 13, 14, 15, 16, 17, 20, 22, 24, —, 100, 121,
10,000.

Solution

The missing number is 31. The terms of the sequence are the representations of 16 to various bases starting with base 16 and proceeding consecutively to base 2.

In general, we avoid "missing term" problems in our series to the extent that we have rejected many similar proposed problems and chose this one only because of the unusual "simplicity" of the solution. This suggests that a valid way of presenting such a problem is to ask for the missing term associated with the "least contrived" or "most simply stated" function instead of just "the missing term" period.

One of our readers, Cornelius Groenewoud of New York, reminded us, of course, that "To the mathematically mature it is well known that there is no such thing as *the* missing number in a specified sequence. It is possible to insert any arbitrary number and to find a formula which will match the given sequence term for term."

We share Mr. Groenewoud's solution with you:

I am enclosing a solution to the problem which will give an arbitrary value c in the 12th position and which gives all the specified numbers in the order given. Thus, for every real number c (and there are a nondenumerable infinity of them) there is a solution of the form indicated below.

There are many other ways of fitting a formula to the 15 positions including the arbitrary c in the 12th position.

The submitted solution was not selected because of its simplicity, but rather to emphasize the wide range of techniques available for formulating the nth term of a sequence of numbers:

Let U_n denote the number in the nth position of the sequence and let $\alpha = [\log_2(16 - n)]$ where $[\ ]$ denotes the greatest integer function. Then:

$$U_n = \frac{\alpha(\alpha - 1)(\alpha - 2)(9 + n)}{6} + \frac{\alpha(\alpha - 2)(\alpha - 3)(n - 3)^2}{2}$$
$$- \frac{(\alpha - 1)(\alpha - 2)(\alpha - 3)(n - 5)^4}{6}$$

$$-\frac{\alpha(\alpha-1)(\alpha-3)}{2}\left\{\left(\frac{c-26}{6}\right)(n-9)^3\right.$$

$$+\left(\frac{26-c}{2}\right)(n-9)^2+\left(\frac{c-20}{3}\right)(n-9)+20\right\}$$

Example: $n = 11$; $16 - n = 5$; $\log_2 5 = 2.32$ $\alpha = [2.32] = 2.0$

$U_{11} = 0 + 0 - 0$

$$-\frac{2.1(-1)}{2}\left\{\frac{c-26}{6}.8+\frac{c-26}{2}.4+\frac{c-20}{3}.2+20\right\}$$

$$=\frac{4}{3}c-\frac{4}{3}.26+52-2c+\frac{2}{3}c-\frac{40}{3}+20$$

$$=72-\frac{144}{3}=72-48=24$$

Example: $n = 12$; $16 - n = 4$; $\log_2 4 = 2.00$ $\alpha = [2.00] = 2.00$

$U_n = 0 + 0 - 0$

$$-\frac{2.1(-1)}{2}\left\{\frac{c-26}{6}.27+\frac{26-c}{2}.9+\frac{c-20}{3}.3+20\right\}$$

$$=\frac{9}{2}c-\frac{9}{2}.26+\frac{9}{2}.26-\frac{9c}{2}+c-20+20=c$$

for *any* real number c.

We thanked our astute reader but pointed out that his solutions involve polynomials in n and the greatest integer function. We hoped he would agree that even this is unduly restrictive.

The ultimate answer to a "missing term" problem was supplied by the professor who, on being asked for the value of U_6, given the values of U_1 through U_5, gave the answer, "one trillion and four." When asked for the sequence he had in mind, he replied simply, "the one with the stated values of U_1 through U_5 and with U_n equal to one trillion and four for n greater than 5." We leave you there.

A FAMILIAR SEQUENCE

What letter follows OTTFFSSE__?

For solutions to problems in this section, turn to page 177.

A MATHEMATICAL QUICKIE

You have just 30 seconds to write down six different six-digit numbers, each of which is divisible by 7, 11, and 13. Done it? Splendid! What's the secret?

A DEARTH OF PRIMES

Find 1000 consecutive nonprime numbers.

CHEAPER BY THE THOUSAND

A noted mathematician was shopping at a hardware store and asked the price of certain articles. The salesman replied, "One would cost 10 cents, eight would cost 10 cents, seventeen would cost 20 cents, one hundred and four cost 30 cents, seven hundred and fifty-six would also cost 30 cents, and one thousand and seventy-two would cost 40 cents." What was the mathematician buying?

TOE-TAC-TIC

The game of reverse tic-tac-toe (known to some as toe-tac-tic) has the same rules as the standard game with one exception. The first player with three markers in a row loses. Can the player with the first move avoid being beaten?

EXCHANGING HANDS

If the hour and minute hands of a watch are inter-changed, how many different possible times could the watch show?

COUNTING PRIMES

How many primes appear in the following infinite sequence, where the digits are arranged in descending order? 9; 98; 987; 9876; . . . . ; 987654321; 9876543219; 98765432198; . . . . . etc.

EFFICIENT WEIGHING

A bricklayer has 8 bricks. Seven of the bricks weigh the same amount and 1 is a little heavier than the others. If the man has a balance scale how can he find the heaviest brick in only 2 weighings?

AN INTERCITY MEETING

A handcar sets out from Chicago to Detroit at an average speed of 10 mph. Four hours later a freight train starts a run from Detroit to Chicago at an average speed of 20 mph. Assuming the rail distance between the two cities is 400 miles, which will be closer to Chicago when they meet?

A FLAG PROBLEM

Suppose three new states, say Guam, Midway, and Puerto Rico were added to the United States. Then the number of states would be 53, a prime. We could no longer design a symmetric field of stars for the flag: true or false?

A UNIQUE DIGITAL PROPERTY

Find N if N^3 and N^4 together contain the 10 digits 0–9 once and only once.

NUMBER CLASSES

The numbers are divided into three groups as follows: 0, 3, 6, 8, 9, . . . , in the first group, 1, 4, 7, 11, 14, . . . , in the second group and 2, 5, 10, 12, 13, . . . , in the third. In which groups would 15, 16, and 17 be placed?

A PECULIAR PROGRESSION

Find three integers in arithmetic progression whose product is prime.

THE CLEVER GRANDFATHER

It is well known that a stopped clock gives the exact time twice a day, while a normally running clock will not be right more than once over a period of months. A clever grandfather adjusted his clock to give the correct time at least twice a day, while running at the normal rate. Assuming he was not able to set it perfectly (a reasonable assumption), how did he do it?

THE GOLD CHAIN

A wizard in numerical analysis has a gold chain with seven links. A lady programmer challenges him to use the chain to buy seven kisses, each kiss to be paid for, separately, with one chain link. What is the smallest number of cuts he will have to make in the chain? What is his sequence of payments?

THE PIGEONHOLES

What is the largest number of pigeonholes that can be occupied by 100 pigeons if each hole is occupied, but no two holes contain the same number of pigeons?

COMPUTING AREA WITHOUT INTEGRATION

An irregular closed curve is drawn on a sheet of paper. What instrument would you require to compute the enclosed area?

Solutions

A Familiar Sequence

N if we consider the letters as the first letters in one, two, three, four, five, six, seven, eight, *n*ine.

A Mathematical Quickie

Any number of the form *abcabc* will do, since *abcabc*/*abc* equals 1001 which equals $7 \cdot 11 \cdot 13$.

A Dearth of Primes

$1001! + 2$; $1001! + 3, \ldots, 1001! + 1001$. This sequence is composite since $n! + A$ is divisible by A as long as $A > 1$ and $< (n + 1)$.

Cheaper by the Thousand

The mathematician was buying numbers (for doors, gates, etc.) and the price was 10 cents per digit.

Toe-tac-tic

Very easily. He takes the center square and then counters each of his opponent's moves by taking the diametrically opposite square. (See "cigar problem" in the Foreword, page vi.)

Exchanging Hands

As the now rapidly rotating hour hand travels from one numeral to the next, it will catch the minute hand at exactly one "possible" time. Thus 12 possible times are achieved during each hour, and in the course of 12 hours, 144 times. But the position in which both hands are pointing upward has been counted twice; the answer is, therefore, 143.

Counting Primes

There are no primes in this series. Consider the first nine numbers, the largest of which is 987,654,321. Eliminating even numbers and those ending in 5 leaves those ending in 9, 7, 3 and 1. But the digital sum of each is either 9 or 6, making them all divisible by 3. Next, any number higher than this number will repeat the same digital sums since the digital sum of this ninth number is 9.

Efficient Weighing

He divides the bricks into 3 groups of 3, 3, and 2 bricks. Then he weighs the 2 sets of 3 bricks against each other. If they balance, then the heavier brick is in the group containing 2 bricks and will be determined by a second weighing. If they do not balance, then he weighs 2 of the group that was heavier on the first weighing. If these 2 balance, then the heavier brick is set aside. If they do not balance, then the scales tell which is the heavier brick.

An Intercity Meeting

When the freight train and the handcar meet, they will be the same distance from Chicago, obviously.

A Flag Problem

False. Seven rows alternately of 8, 7, 8, 7, 8, 7, and 8 stars would do very nicely.

A Unique Digital Property

If $N \geq 22$, N^3 will contain at least 5 digits, N^4 at least 6, so that they will contain at least 11 digits together. Likewise if $N \leq 17$, N^3 and N^4 will contain at most 9 digits together. Hence $N = 18, 19, 20,$ or 21. Trying each in turn, we find only $18^3 = 5832$, $18^4 = 104976$. Therefore $N = 18$.

Number Classes

The first group consists of numbers written only with curves, the second group consists of numbers written only with straight lines, and the third group consists of numbers written with both straight lines and curves. Therefore 15 and 16 would be in the third group and 17 in the second.

A Peculiar Progression

By definition of "prime," two of the integers must be of unit absolute value. Hence the three integers are -3, -1, $+1$.

The Clever Grandfather

He adjusted the clock to run backwards.

The Gold Chain

One cut in the third link will allow two links to be swapped for a kiss and a link on the second transaction, and three links for a kiss and two links on the third, and so on.

The Pigeonholes

$1 + 2 + 3 + \cdots + 13 = 91$, while $1 + 2 + 3 + \cdots + 14 = 105$. With a little cogitation, one sees that for any number of pigeons from 91 to 104, the answer is 13 pigeonholes.

Computing Area without Integration

A sensitive scale will do. Measure the sides of the sheet and determine the area of the sheet. Next weigh the sheet. Finally, cut out the area in question, weigh it, and compare its weight with that of the sheet. The areas, of course, will be in the same ratio.

7

PERMUTATIONS, PARTITIONS, AND PRIMES

ЛЛЛЛЛЛЛЛЛЛЛЛЛЛЛЛЛЛЛ

Assorted Number Theory Problems

Splitting a Number (SOLUTION, Page 184)
A Power of a Power (SOLUTION, Page 186)
An Impossible Partition (SOLUTION, Page 188)
Odd Lots (SOLUTION, Page 190)
The Transferred Digit · A Question of Primality · Primes and Squares · A Divisibility Problem · A Matter of Address · Points on a Circle · An Exponential Equation · The Picture and the Frame · The Rose Garden · Ice Cream Sales · Variation on an Old Theme · Still Another Age Problem · Driving Economy · Find the Base · Find the Power · Find the Digits · Square Pairs · The Left-handed Complement · Ones and Twos · The First Day of the Century · A Large Number · A Problem of Remainders · A Problem of Divisors · Never a Square

SOLUTIONS: Page 212

SPLITTING A NUMBER

If a certain six-digit number is split into two parts, one constituting the first three digits and the other the last three digits, and the two parts are added and the resulting sum squared, it is found that the product is the original six-digit number. What is the number?

Solution

$$998,001 = (998 + 001)^2$$

There is one other six-digit number with similar properties: $494,209 = (494 + 209)^2$. If, as we had originally considered, we had added a restriction to the statement of the problem that neither of the two parts begin with a zero, we would have (1) ruled out our published answer of 998,001, (2) printed the only other answer of 494,209, and (3) avoided answering considerable correspondence on the problem.

"So what's wrong with 494,209?" asked a reader in Alamogordo, New Mexico. "In fact I think it's even better than your solution because each part is a three-digit number in itself."

From a naval commander: "The reason I am writing is to fuss about the answer to problem #136. 998,001 is not the only answer. 494,209 is also a solution."

But if we had not goofed, we might never have met the gentleman from Canada, Dr. Rudolf G. de Buda, who kindly sent us an excellent direct proof showing that 998,001 and 494,209 are the only two answers to the problem. We hasten to share Dr. de Buda's proof with you:

Let the number be $1000a + b$ with $c = a + b$. Then $1000a + b = c^2$ gives $a = \dfrac{c(c - 1)}{999}$. Hence, between them, c and $c - 1$ contain the factors of 999, i.e., 37 and 27. $c - 1 \neq 999$, since a has 3 digits. But $c = 999$ provides one solution, namely 998,001. Otherwise c and $c - 1$ each contain one of the factors. Either $c = 37p$ and $c - 1 = 27q$ or $c = 27q$ and $c - 1 = 37q$. Solving these linear diophantine equations, $c = 703 + 999k$ or $c = 297 + 999m$. Only the former (with $k = 0$) is the square root of a six-digit number giving rise to the other solution, 494,209.

A POWER OF A POWER

What is the rightmost digit of 7^{7^7}?

Solution

The rightmost digit of 7^N is 1, 7, 9, or 3 according as $N \equiv 0$, 1, 2, or 3 (mod 4). Since $7^7 \equiv 3^7 \equiv 3$ (mod 4), the rightmost digit of 7^{7^7} is 3.

A number of our readers misinterpreted this number theory problem and insisted that the answer was 7 not 3. "Are you sure the answer is not 7?" asked an Alaskan reader. "May I be the 7^{7^7} person to suggest that the answer given to your puzzle is incorrect?" wrote an engineer in New York. "Please send me a letter indicating that you have erred in the listing of the answer to your last puzzle," insisted a reader in Illinois.

Though their analyses were impeccable, our readers misread the problem and needed only to remember that the standard mathematical convention is to read a^{b^c} as $a^{(b^c)}$, not $(a^b)^c$, and the stated value of three is, therefore, correct. In other words $7^{7^7} = 7^{(7^7)}$ not 7^{49}.

AN IMPOSSIBLE PARTITION

Can 19^{19} be represented as the sum of a cube and a fourth power?

Solution

No. From the multiplication table modulo 13 one sees that $X^3 \equiv 0$, 1, 5, 8, or 12 and $Y^4 \equiv 0, 1, 3,$ or 9. Thus $X^3 + Y^4$ can be congruent to anything except 7 (mod 13). But $19^{19} \equiv 6^{19} \equiv 7$ (mod 13).

This one's a bit difficult for those who haven't had number theory. May we refer you to *An Introduction to the Theory of Numbers*, G. H. Hardy and E. M. Wright, Third Edition, Oxford Clarendon Press, 1954, page 49. We referred many of our readers. Among them was an American working in the Netherlands who wrote:

> Enough is enough. . . . I don't mind if you have a tricky one that hits below the belt, but I can't even read the solution to this one. . . . The least you could do in the solution is to give a reference that would explain this weird mathematics that has 19^{19} identical to 6^{19}, 6^7, and 7 (mod 13). . . . Now come on, give me a text reference to this notation!

From an Atlanta attorney came the following case against the problem:

> I disagree and present the following solution:
>
> Where $X^3 + Y^4 = 19^{19}$
>
> Let $X = \sqrt[3]{\dfrac{19^{19}}{2}}$ and $Y = \sqrt[4]{\dfrac{19^{19}}{2}}$

We raised an objection to the counselor's case. The problem implicitly requires a solution in terms of integers, i.e., it is a "diophantine equation," since without this restriction there are obviously infinitely many solutions. However, we're not certain this objection would be sustained in court.

ODD LOTS

A set of items sells for $1122.00, and another set of like items sells for $2210.00. What is the cost of each item?

Solution

$17.00. $1122 = N_1 \cdot C$, $2210 = N_2 \cdot C$ so that C divides 1122 and 1105, and hence C divides the greatest common divisor of these number. The g.c.d. is 17, a prime, so that $C = 17$ with $N_1 = 66$, $N_2 = 130$.

Don't you believe it. Our solution, that is. We slipped on this problem, and we heard about it from all corners. Our first error was in saying that C divides 1105. This lost the valid solution of $34, as pointed out by numerous correspondents from California to Massachusetts. Even if we had specified that each item cost an integral number of dollars, there would have been four solutions said E. Gerding of Phoenix, Arizona, to wit, $1, $2, $17, and $34. But we hadn't so specified, so that, as D. Feign of Santa Ana, California, advised, there are 24 distinct solutions, representing the 24 integral divisors of 3,400 (cents). The problem and its response taught us to pay closer attention to problem proposals with an eye both to accuracy of solution and to sufficiently restrictive wording of the problem itself.

As a belated penance, we print all 24 solutions: $34, $17, $8.50, $6.80, $4.25, $3.40, $2, $1.70, $1.36, $1, 85 cents, 68 cents, 50 cents, 40 cents, 34 cents, 25 cents, 20 cents, 17 cents, 10 cents, 8 cents, 5 cents, 4 cents, 2 cents, and 1 cent.

THE TRANSFERRED DIGIT

Find the smallest integer which is such that if the digit on the extreme left is transferred to the extreme right, the new number is three and a half times the original number.

For solutions to problems in this section, turn to page 212.

A QUESTION OF PRIMALITY

For what values of n is $11 \times 14^n + 1$ prime?

PRIMES AND SQUARES

The sum and difference of two squares may be primes: $4 - 1 = 3$ and $4 + 1 = 5$; $9 - 4 = 5$ and $9 + 4 = 13$, etc. Can the sum and difference of two primes be squares? If so, for how many different primes is this possible?

A DIVISIBILITY PROBLEM

How many nine-digit numbers are divisible by 11, no digit equal to zero and no two digits alike?

A MATTER OF ADDRESS

My house is on a road where the numbers run 1, 2, 3, 4,
. . . consecutively. By a curious coincidence, the sum
of all house numbers less than mine is the same as the
sum of all house numbers greater than mine. What is
my number and how many houses are there on my road
if my house number is in the thirties?

POINTS ON A CIRCLE

There are n points on a circle. A straight line segment is drawn between each pair of points. How many intersections are there within the circle if no 3 lines are concurrent?

AN EXPONENTIAL EQUATION

Find unequal rational numbers, a, b, (other than 2 and 4) such that $a^b = b^a$.

THE PICTURE AND THE FRAME

A rectangular picture, each of whose dimensions is an integral number of inches, has an ordinary rectangular frame 1 inch wide. Find the dimensions of the picture if the area of the picture and the area of the frame are equal.

THE ROSE GARDEN

Mr. Perkins decided to redesign his rectangular rose bed into the shape of a right-angled triangle. The existing bed measured 24 by 35 feet. He discovered that he could make any one of three different right-triangular beds, each equal in area to the existing bed and each having sides of an integral number of feet. As it was his custom to fence his beds, he naturally chose the bed with the smallest perimeter. What were the dimensions, in feet, of the new bed?

ICE CREAM SALES

1960 and 1961 were bad years for ice cream sales but 1962 was very good. An accountant was looking at the tonnage sold in each year and noticed that the digital sum of the tonnage sold in 1962 was three times as much as the digital sum of the tonnage sold in 1961. Moreover, if the amount sold in 1960 (346 tons) was added to the 1961 tonnage, this total was less than the total tonnage sold in 1962 by the digital sum of the tonnage sold in that same year. Just how many more tons of ice cream were sold in 1962 than in the previous year?

VARIATION ON AN OLD THEME

Choose a number less than 40. Add 1 and multiply by the first number. Add 41 and square the result. Divide by 12 and the remainder is 1. Why?

STILL ANOTHER AGE PROBLEM

Three times Dick's age plus Tom's age equals twice Harry's age. Double the cube of Harry's age is equal to three times the cube of Dick's age added to the cube of Tom's age. Their respective ages are relatively prime to each other. How old are they?

DRIVING ECONOMY

An astute mathematician drives 21 miles round trip to work each day. On the way he passes a gas station which advertises free gas if the price at which the pump stops when filling the tank consists of repetitive digits, i.e., $1.11, $2.22, $3.33, . . . , $9.99. Gas costs 30 cents per gallon and our mathematician knows his car delivers exactly 15 miles per gallon. Considering no additional driving, he computes that once he fills his gas tank at the station he can get all his gas free. The station is an integral number of miles from his home. Where is it with respect to his home?

FIND THE BASE

Find the digital base in which the number seven thousand, six hundred and forty-two is represented by the symbol 1234.

FIND THE POWER

What is the largest power of 7 that will divide 1000! ?

FIND THE DIGITS

Find a two-digit number which is a factor of the sum of the cubes of its digits, while the reverse of the number is a factor of the sum of the fourth powers of the digits.

SQUARE PAIRS

There are pairs of numbers whose sum and product are perfect squares. For instance $5 + 20 = 25$ and $5 \times 20 = 100$. If the smaller number of such a pair is 1090, what is the smallest possible value of the other?

THE LEFT-HANDED COMPLEMENT

It is reckoned that $1/x$ of the population of a certain country is exclusively left-handed, and that $1/y$ of the population is ambidextrous, x and y being integers. The number of exclusively left-handed persons exceeds that of the ambidextrous ones by the smallest possible number consistent with the fact that both groups together constitute $\frac{1}{91}$ of the population. What fraction of the population is exclusively left-handed and what fraction is ambidextrous?

ONES AND TWOS

The smaller of two consecutive integers is divisible by 23 and the larger by 29. Find the smallest pair of such numbers with the property that they both contain only the digits 1 and 2.

THE FIRST DAY OF THE CENTURY

On what days of the week can the first day of a century fall? (The first day of the twentieth century was Jan. 1, 1901.)

A LARGE NUMBER

What are the last three digits of the number 7^{9999}?

A PROBLEM OF REMAINDERS

What number, if divided by 10, leaves a remainder of 9; divided by 9 leaves a remainder of 8; divided by 8 leaves a remainder of 7; . . . ; divided by 2 leaves a remainder of 1. One answer is 14,622,042,959. Find a smaller solution.

A PROBLEM OF DIVISORS

Find the smallest number with 28 divisors.

NEVER A SQUARE

If P_n denotes the nth prime, show that $P_1 P_2 \cdots P_n + 1$ is not a perfect square.

Solutions

The Transferred Digit

Let the number be $abc \cdots z$. Then: $2bc \cdots za = 7abc \cdots z$, and
$b(2 \cdot 10^k - 7 \cdot 10^{k-1}) + c(2 \cdot 10^{k-1} - 7 \cdot 10^{k-2}) + \cdots z(20 - 7)$

$$+ b(13 \cdot 10^{k-1}) + c(13 \cdot 10^{k-2}) + \cdots + z(13)$$
$$= a(7 \cdot 10^k - 2)$$

Since the prime 13 divides the left member it must also divide the right, and in particular, the second factor. Testing in turn the values 68, 698, 6998, 69998, and 699998, the latter is the first to be divisible by 13. Thus $k = 5$ and our number has 6 digits $abcdef$. Also $130,000b + 13,000c + 1,300d + 130e + 13f = 699,998a$ or $bcdef = 53,846a$. Hence $a = 1$ and the desired number is 153,846.

A Question of Primality

If n is odd, 14^n ends in 4. So does 11×14^n. Therefore, $11 \times 14^n + 1$ ends in 5 and hence is divisible by 5. If n is even, say equal to $2m$, $11 \times 14^n + 1$ is equal to $11 \times 196^m + 1$. Since $196 = 3 \times 65 + 1$, $(3 \times 65 + 1)^m$ will always leave a remainder of 1 when divided by 3. Therefore $11 \times 196^m + 1$ is equal to $11(3k + 1) + 1$ or $33k + 12$. Hence when n is even, $11 \times 14^n + 1$ is always divisible by 3. Therefore, $11 \times 14^n + 1$ is never a prime, being a multiple of 5 or 3 according as n is odd or even.

Primes and Squares

Subtracting $p - q = y^2$ from $p + q = x^2$, where p and q are primes, we have: $2q = x^2 - y^2 \neq 4n + 2$. Then $q \neq 2n + 1$. Therefore, $q = p = 2$ is the only solution.

A Divisibility Problem

A number is divisible by 11 if and only if the sum of its 1st, 3rd, 5th, etc., digit differs by a multiple of 11 from the sum of its 2nd, 4th, 6th, etc., digits. In the problem at hand the sum of all 9 digits is 45. The two sums are, therefore, 28 and 17 or 39 and 6 and the latter may be quickly ruled out. By trial and error one finds that there are nine ways in which 4 digits can sum to 17 and two ways in which they can sum to 28. Thus there are 11 choices for the even-place digits. These may be permuted in 4! ways and for each choice the odd-place digits may be permuted in 5! ways. The solution is, therefore,

$$11 \times 24 \times 120 = 31,680$$

A Matter of Address

Let x be the house number and y the number of houses on the road. Then:

$$1 + 2 + \cdots + x - 1 = (x + 1) + (x + 2) + \cdots + y$$

$$\frac{x(x - 1)}{2} = \frac{y(y + 1)}{2} - \frac{x(x + 1)}{2}$$

$$2x^2 = y^2 + y$$

Solving as a quadratic in y,

$$y = \frac{1 \pm \sqrt{1 + 8x^2}}{2}.$$

In order for y to be an integer, $1 + 8x^2$ must be a perfect square. For $30 \leq x \leq 39$ this holds only when $x = 35$ and $y = 49$.

Points on a Circle

Let A, B, C, D be any 4 of the points. Let them so be ordered that $ABCD$ is a polygon. Then the lines AC and BD are uniquely determined and they form one intersection inside the circle. To each set of 4 points there corresponds a unique intersection within the circle. Hence there are

$$C_n^4, \quad \text{or} \quad \frac{n(n - 1)(n - 2)(n - 3)}{4 \cdot 3 \cdot 2 \cdot 1}$$

such intersections.

An Exponential Equation

Let $b = ra$
Then:

$$a^{ra} = (ra)^a$$
$$a^r = ra$$
$$a^{r-1} = r$$
$$a = r^{\frac{1}{r-1}}$$

Assuming r is rational, a will be rational whenever $\dfrac{1}{r - 1}$ is an integer,

i.e., when $r = 1 + \dfrac{1}{k}$.

Thus

$$a = \left(1 + \frac{1}{k}\right)^k; \quad b = \left(1 + \frac{1}{k}\right)^{k+1}$$

satisfies the given equation for $k = \pm 1, \pm 2, \pm 3, \ldots$.

The Picture and the Frame

Denoting width and length a and b, we obtain the equation $a = (2b+4)/(b-2)$, yielding 2 integral solutions: 3×10 or 4×6.

The Rose Garden

The problem is one of finding three right triangles of different dimensions but of the same area, viz., 840 square feet. There are just three such triangles and their dimensions are (1) 42, 40, 58; (2) 70, 24, 74; and (3) 112, 15, 113. Number 1 has the smallest perimeter and is, therefore, the one Perkins chose.

Ice Cream Sales

361 tons. If we call the tonnages of 1961 and 1962 A and B respectively, and their digital sums x and $3x$, we have:

$B - A = 346 + 3x$; digitally $2x = 13 + 3x$. The only possible answer is $x = 5$. Hence $B - A = 361$. (It is not possible to be sure of the amounts sold in 1961 or 1962, but 361 is unique).

Variation on an Old Theme

It's a well-known fact that $n^2 + n + 41$ is prime for $n < 40$. Since it is greater than 3, such a number is of the form $6k \pm 1$, which is one more than a multiple of 12 when squared.

Only after this problem appeared in the Problematical Recreations series were we advised that the proposition is true for all integers rather than only for those less than 40. J. N. A. Hawkins of Pacific Palisades, California, furnished this neat proof:

$$(n(n + 1) + 41)^2 \equiv (n^2 + n + 5)^2 \equiv n^4 + 2n^3 + 11n^2 + 10n + 25$$
$$\equiv n^4 + 2n^3 - n^2 - 2n + 1$$
$$\equiv n(n - 1)(n + 1)(n + 2) = 1 \pmod{12}$$

but the product of four consecutive integers has some factor divisible by 3 and some factor divisible by 4. Hence

$$(n(n + 1) + 41)^2 \equiv 1 \pmod{12}$$

Still Another Age Problem

Let $x = $ Dick's age, $y = $ Tom's age, $z = $ Harry's age. Then $3x + y = 2z$. $2z^2 = 3x^3 + y^3 = 3x^3 + (2z - 3x)^3 = 3x^3 + 8z^3 - 36z^2x + 54zx^2 - 27x^3$ or $1w^3 - 6w^2 + 9w - 4 = 0$ with $w = z/x$.

Since x and z are integers, w is rational. The only possible positive rational solutions of 1 are $w = 1, 2, 4$. $w = 2$ does not work. $w = 1$ satisfies, but this means $z = x$, impossible since x and z are relatively prime. Hence, $w = 4 = z/x$. $w = 4$ satisfies Eq. 1. Since x and z are relatively prime, $x = 1$, $z = 4$, and $y = 5$.

Driving Economy

$1.11 worth of gas is 3.7 gallons, which will yield 55.5 miles. Let the N's be integers and X be the distance from his home to the station. $(55.5)N$ is not divisible by 21 for $N \leq 9$(maximum sale $9.99).

However,
$$N_1(55.5) = N_2(21) + 2X$$
$$N_3(55.5) = N_4(21) + (21 - 2X)$$

is a valid pair of equations, considering that he can stop, either on his way home, or on his way to work (and since $55.5N$ is not divisible by 21, it must be some combination of these alternatives that he uses).

$$\text{Adding} \quad (N_1 + N_3)55.5 = (N_2 + N_4 + 1)21$$

$(N_1 + N_3)$ must not exceed 18 because of the limit of $9 \times \$1.11$, so a very simple process gives $N_1 + N_3 = 14$ and $N_2 + N_4 = 36$.

Integral values of N_1 and $N_3 \leq 9$ show N_1 and $N_3 = 7$ and 7, 6 and 8, or 5 and 9, yielding in turn (from pricing and mileage) N_2 and $N_4 = 18$ and 18, 15 and 21, and 13 and 33. Solving for X, we find it equals 1.5 or 9, 2.25 or 8.25, or 5.25. Since X must be an integer the station is 9 miles from his home.

Find the Base

Letting N denote the base, $N^3 + 2N^2 + 3N + 4 = 7,642$ or $N^3 + 2N^2 + 3N - 7,638 = 0$. N is evidently larger than 10 and must be a factor of $7,368 = 2.3.19.67$. Trying 19 first, we find it satisfies the equation. 19 is, therefore, the desired base.

Find the Power

The product $1.2.3. \ldots . 1000$ contains $[\frac{1000}{7}]$ factors which are multiples of 7, or 142. Of these, $[\frac{1000}{49}] = 20$ are multiples of 49 and therefore contain an additional seven. Finally, of these, $[\frac{1000}{343}] = 2$ contain a third seven. The largest power of 7 which divides 1000! is, therefore, the $142 + 20 + 2 =$ the 164th power.

Find the Digits

The cases where $A^3 + B^3$ has a factor $10A + B$ are given by:

$$2^3 + 4^3 = 3 \times 24$$
$$2^3 + 7^3 = 13 \times 27$$
$$3^3 + 7^3 = 10 \times 37$$
$$4^3 + 8^3 = 12 \times 48$$

Forming $A^4 + B^4$ for each of these we find only the third satisfies the requirements. $3^4 + 7^4 = 34 \times 73$. Therefore $A = 3$, $B = 7$.

Square Pairs

We have $1090 + x = A^2$ and $1090x = B^2$. Since 1090 has no square factor > 1, 1090 must be a factor of x; in fact x must be of the form $1090c^2$. Thus $A^2 = 1090 \, (c^2 + 1)$, and by similar reasoning 1090 is a factor of $c^2 + 1$. The smallest value of c^2 which makes $c^2 + 1$ divisible by 1090 is 1089 which, fortunately, happens to be a perfect square, i.e., 33^2. Hence $x = 1090 \, (1089) = 1,187,010$.

The Left-handed Complement

We must solve $\frac{1}{91} = 1/x + 1/y$. If $91 = abc$, then $x = ac(a + b)$ and $y = bc(a + b)$. This leads to four sets of solutions, of which the set we want is $\frac{1}{110}$ and $\frac{1}{910}$, which are the fractions of exclusively left-handed and ambidextrous people respectively.

Ones and Twos

Let $N = 23a$, $N + 1 = 29b$. Then $29b = 23a + 1$, from which a is of the form $29k + 5$ and b is of the form $23k + 4$. Obviously N ends in 1, so that $23a$ ends in 1 and a ends in 7. Therefore k must end in 8. We find $k = 18$, $a = 527$, $b = 418$, so that the two numbers are 12121 and 12122.

The First Day of the Century

If D is the number of the day in year Y,

$$X = D + Y + \left[\frac{Y-1}{4}\right] - \left[\frac{Y-1}{100}\right] + \left[\frac{Y-1}{400}\right]$$

(mod 7) is the day of the week D falls on (Sunday = 1, Monday = 2, etc.). If $D = 1$, $Y = 100N + 1$, this becomes

$$X = 124N + 2 + \left[\frac{N}{4}\right] \equiv 5N + 2 + \left[\frac{N}{4}\right] \pmod{7}$$

This congruence has only the solutions 0, 2, 3, and 5. Therefore the first day of a century can fall only on Monday, Tuesday, Thursday, or Saturday. The twenty-first century will start on a Monday.

A Large Number

Notice that the given number is equivalent to seven raised to ten thousand divided by seven and that ten thousand is equal to four times twenty-five hundred; then introducing a suitable variable, k, one ends with a two-term number. The first term is equal to $10^3(k - 1)$

divided by seven and the second term, 143, is the solution to the problem.

A Problem of Remainders

3,628,799 will do. Also, since the l.c.m. of 1, 2, . . . , 10 is 2520, 2519 is the smallest positive solution.

A Problem of Divisors

Since $28 = 2 \cdot 2 \cdot 7$, the number must be of the form $2^a 3^b 5^c$ where $(a + 1)(b + 1)(c + 1) = 28$. The smallest such number is $2^6 \cdot 3 \cdot 5$, or 960.

Never a Square

Since $P_1 P_2 \cdots P_n + 1$ will leave a remainder of 3 on division by 4, while any square leaves a remainder of 0 or 1, it follows that $P_1 P_2 \cdots P_n + 1$ is never a square.

A CATALOG OF SELECTED
DOVER BOOKS
IN SCIENCE AND MATHEMATICS

Astronomy

CHARIOTS FOR APOLLO: The NASA History of Manned Lunar Spacecraft to 1969, Courtney G. Brooks, James M. Grimwood, and Loyd S. Swenson, Jr. This illustrated history by a trio of experts is the definitive reference on the Apollo spacecraft and lunar modules. It traces the vehicles' design, development, and operation in space. More than 100 photographs and illustrations. 576pp. 6 3/4 x 9 1/4. 0-486-46756-2

EXPLORING THE MOON THROUGH BINOCULARS AND SMALL TELESCOPES, Ernest H. Cherrington, Jr. Informative, profusely illustrated guide to locating and identifying craters, rills, seas, mountains, other lunar features. Newly revised and updated with special section of new photos. Over 100 photos and diagrams. 240pp. 8 1/4 x 11. 0-486-24491-1

WHERE NO MAN HAS GONE BEFORE: A History of NASA's Apollo Lunar Expeditions, William David Compton. Introduction by Paul Dickson. This official NASA history traces behind-the-scenes conflicts and cooperation between scientists and engineers. The first half concerns preparations for the Moon landings, and the second half documents the flights that followed Apollo 11. 1989 edition. 432pp. 7 x 10.
0-486-47888-2

APOLLO EXPEDITIONS TO THE MOON: The NASA History, Edited by Edgar M. Cortright. Official NASA publication marks the 40th anniversary of the first lunar landing and features essays by project participants recalling engineering and administrative challenges. Accessible, jargon-free accounts, highlighted by numerous illustrations. 336pp. 8 3/8 x 10 7/8. 0-486-47175-6

ON MARS: Exploration of the Red Planet, 1958-1978—The NASA History, Edward Clinton Ezell and Linda Neuman Ezell. NASA's official history chronicles the start of our explorations of our planetary neighbor. It recounts cooperation among government, industry, and academia, and it features dozens of photos from Viking cameras. 560pp. 6 3/4 x 9 1/4. 0-486-46757-0

ARISTARCHUS OF SAMOS: The Ancient Copernicus, Sir Thomas Heath. Heath's history of astronomy ranges from Homer and Hesiod to Aristarchus and includes quotes from numerous thinkers, compilers, and scholasticists from Thales and Anaximander through Pythagoras, Plato, Aristotle, and Heraclides. 34 figures. 448pp. 5 3/8 x 8 1/2.
0-486-43886-4

AN INTRODUCTION TO CELESTIAL MECHANICS, Forest Ray Moulton. Classic text still unsurpassed in presentation of fundamental principles. Covers rectilinear motion, central forces, problems of two and three bodies, much more. Includes over 200 problems, some with answers. 437pp. 5 3/8 x 8 1/2. 0-486-64687-4

BEYOND THE ATMOSPHERE: Early Years of Space Science, Homer E. Newell. This exciting survey is the work of a top NASA administrator who chronicles technological advances, the relationship of space science to general science, and the space program's social, political, and economic contexts. 528pp. 6 3/4 x 9 1/4.
0-486-47464-X

STAR LORE: Myths, Legends, and Facts, William Tyler Olcott. Captivating retellings of the origins and histories of ancient star groups include Pegasus, Ursa Major, Pleiades, signs of the zodiac, and other constellations. "Classic." – *Sky & Telescope.* 58 illustrations. 544pp. 5 3/8 x 8 1/2. 0-486-43581-4

A COMPLETE MANUAL OF AMATEUR ASTRONOMY: Tools and Techniques for Astronomical Observations, P. Clay Sherrod with Thomas L. Koed. Concise, highly readable book discusses the selection, set-up, and maintenance of a telescope; amateur studies of the sun; lunar topography and occultations; and more. 124 figures. 26 halftones. 37 tables. 335pp. 6 1/2 x 9 1/4. 0-486-42820-6

Browse over 9,000 books at www.doverpublications.com

Chemistry

MOLECULAR COLLISION THEORY, M. S. Child. This high-level monograph offers an analytical treatment of classical scattering by a central force, quantum scattering by a central force, elastic scattering phase shifts, and semi-classical elastic scattering. 1974 edition. 310pp. 5 3/8 x 8 1/2.   0-486-69437-2

HANDBOOK OF COMPUTATIONAL QUANTUM CHEMISTRY, David B. Cook. This comprehensive text provides upper-level undergraduates and graduate students with an accessible introduction to the implementation of quantum ideas in molecular modeling, exploring practical applications alongside theoretical explanations. 1998 edition. 832pp. 5 3/8 x 8 1/2.   0-486-44307-8

RADIOACTIVE SUBSTANCES, Marie Curie. The celebrated scientist's thesis, which directly preceded her 1903 Nobel Prize, discusses establishing atomic character of radioactivity; extraction from pitchblende of polonium and radium; isolation of pure radium chloride; more. 96pp. 5 3/8 x 8 1/2.   0-486-42550-9

CHEMICAL MAGIC, Leonard A. Ford. Classic guide provides intriguing entertainment while elucidating sound scientific principles, with more than 100 unusual stunts: cold fire, dust explosions, a nylon rope trick, a disappearing beaker, much more. 128pp. 5 3/8 x 8 1/2.   0-486-67628-5

ALCHEMY, E. J. Holmyard. Classic study by noted authority covers 2,000 years of alchemical history: religious, mystical overtones; apparatus; signs, symbols, and secret terms; advent of scientific method, much more. Illustrated. 320pp. 5 3/8 x 8 1/2.

0-486-26298-7

CHEMICAL KINETICS AND REACTION DYNAMICS, Paul L. Houston. This text teaches the principles underlying modern chemical kinetics in a clear, direct fashion, using several examples to enhance basic understanding. Solutions to selected problems. 2001 edition. 352pp. 8 3/8 x 11.   0-486-45334-0

PROBLEMS AND SOLUTIONS IN QUANTUM CHEMISTRY AND PHYSICS, Charles S. Johnson and Lee G. Pedersen. Unusually varied problems, with detailed solutions, cover of quantum mechanics, wave mechanics, angular momentum, molecular spectroscopy, scattering theory, more. 280 problems, plus 139 supplementary exercises. 430pp. 6 1/2 x 9 1/4.   0-486-65236-X

ELEMENTS OF CHEMISTRY, Antoine Lavoisier. Monumental classic by the founder of modern chemistry features first explicit statement of law of conservation of matter in chemical change, and more. Facsimile reprint of original (1790) Kerr translation. 539pp. 5 3/8 x 8 1/2.   0-486-64624-6

MAGNETISM AND TRANSITION METAL COMPLEXES, F. E. Mabbs and D. J. Machin. A detailed view of the calculation methods involved in the magnetic properties of transition metal complexes, this volume offers sufficient background for original work in the field. 1973 edition. 240pp. 5 3/8 x 8 1/2.   0-486-46284-6

GENERAL CHEMISTRY, Linus Pauling. Revised third edition of classic first-year text by Nobel laureate. Atomic and molecular structure, quantum mechanics, statistical mechanics, thermodynamics correlated with descriptive chemistry. Problems. 992pp. 5 3/8 x 8 1/2.   0-486-65622-5

ELECTROLYTE SOLUTIONS: Second Revised Edition, R. A. Robinson and R. H. Stokes. Classic text deals primarily with measurement, interpretation of conductance, chemical potential, and diffusion in electrolyte solutions. Detailed theoretical interpretations, plus extensive tables of thermodynamic and transport properties. 1970 edition. 590pp. 5 3/8 x 8 1/2.   0-486-42225-9

Browse over 9,000 books at www.doverpublications.com

Engineering

FUNDAMENTALS OF ASTRODYNAMICS, Roger R. Bate, Donald D. Mueller, and Jerry E. White. Teaching text developed by U.S. Air Force Academy develops the basic two-body and n-body equations of motion; orbit determination; classical orbital elements, coordinate transformations; differential correction; more. 1971 edition. 455pp. 5 3/8 x 8 1/2. 0-486-60061-0

INTRODUCTION TO CONTINUUM MECHANICS FOR ENGINEERS: Revised Edition, Ray M. Bowen. This self-contained text introduces classical continuum models within a modern framework. Its numerous exercises illustrate the governing principles, linearizations, and other approximations that constitute classical continuum models. 2007 edition. 320pp. 6 1/8 x 9 1/4. 0-486-47460-7

ENGINEERING MECHANICS FOR STRUCTURES, Louis L. Bucciarelli. This text explores the mechanics of solids and statics as well as the strength of materials and elasticity theory. Its many design exercises encourage creative initiative and systems thinking. 2009 edition. 320pp. 6 1/8 x 9 1/4. 0-486-46855-0

FEEDBACK CONTROL THEORY, John C. Doyle, Bruce A. Francis and Allen R. Tannenbaum. This excellent introduction to feedback control system design offers a theoretical approach that captures the essential issues and can be applied to a wide range of practical problems. 1992 edition. 224pp. 6 1/2 x 9 1/4. 0-486-46933-6

THE FORCES OF MATTER, Michael Faraday. These lectures by a famous inventor offer an easy-to-understand introduction to the interactions of the universe's physical forces. Six essays explore gravitation, cohesion, chemical affinity, heat, magnetism, and electricity. 1993 edition. 96pp. 5 3/8 x 8 1/2. 0-486-47482-8

DYNAMICS, Lawrence E. Goodman and William H. Warner. Beginning engineering text introduces calculus of vectors, particle motion, dynamics of particle systems and plane rigid bodies, technical applications in plane motions, and more. Exercises and answers in every chapter. 619pp. 5 3/8 x 8 1/2. 0-486-42006-X

ADAPTIVE FILTERING PREDICTION AND CONTROL, Graham C. Goodwin and Kwai Sang Sin. This unified survey focuses on linear discrete-time systems and explores natural extensions to nonlinear systems. It emphasizes discrete-time systems, summarizing theoretical and practical aspects of a large class of adaptive algorithms. 1984 edition. 560pp. 6 1/2 x 9 1/4. 0-486-46932-8

INDUCTANCE CALCULATIONS, Frederick W. Grover. This authoritative reference enables the design of virtually every type of inductor. It features a single simple formula for each type of inductor, together with tables containing essential numerical factors. 1946 edition. 304pp. 5 3/8 x 8 1/2. 0-486-47440-2

THERMODYNAMICS: Foundations and Applications, Elias P. Gyftopoulos and Gian Paolo Beretta. Designed by two MIT professors, this authoritative text discusses basic concepts and applications in detail, emphasizing generality, definitions, and logical consistency. More than 300 solved problems cover realistic energy systems and processes. 800pp. 6 1/8 x 9 1/4. 0-486-43932-1

THE FINITE ELEMENT METHOD: Linear Static and Dynamic Finite Element Analysis, Thomas J. R. Hughes. Text for students without in-depth mathematical training, this text includes a comprehensive presentation and analysis of algorithms of time-dependent phenomena plus beam, plate, and shell theories. Solution guide available upon request. 672pp. 6 1/2 x 9 1/4. 0-486-41181-8

Browse over 9,000 books at www.doverpublications.com

HELICOPTER THEORY, Wayne Johnson. Monumental engineering text covers vertical flight, forward flight, performance, mathematics of rotating systems, rotary wing dynamics and aerodynamics, aeroelasticity, stability and control, stall, noise, and more. 189 illustrations. 1980 edition. 1089pp. 5 5/8 x 8 1/4.   0-486-68230-7

MATHEMATICAL HANDBOOK FOR SCIENTISTS AND ENGINEERS: Definitions, Theorems, and Formulas for Reference and Review, Granino A. Korn and Theresa M. Korn. Convenient access to information from every area of mathematics: Fourier transforms, Z transforms, linear and nonlinear programming, calculus of variations, random-process theory, special functions, combinatorial analysis, game theory, much more. 1152pp. 5 3/8 x 8 1/2.   0-486-41147-8

A HEAT TRANSFER TEXTBOOK: Fourth Edition, John H. Lienhard V and John H. Lienhard IV. This introduction to heat and mass transfer for engineering students features worked examples and end-of-chapter exercises. Worked examples and end-of-chapter exercises appear throughout the book, along with well-drawn, illuminating figures. 768pp. 7 x 9 1/4.   0-486-47931-5

BASIC ELECTRICITY, U.S. Bureau of Naval Personnel. Originally a training course; best nontechnical coverage. Topics include batteries, circuits, conductors, AC and DC, inductance and capacitance, generators, motors, transformers, amplifiers, etc. Many questions with answers. 349 illustrations. 1969 edition. 448pp. 6 1/2 x 9 1/4.
0-486-20973-3

BASIC ELECTRONICS, U.S. Bureau of Naval Personnel. Clear, well-illustrated introduction to electronic equipment covers numerous essential topics: electron tubes, semiconductors, electronic power supplies, tuned circuits, amplifiers, receivers, ranging and navigation systems, computers, antennas, more. 560 illustrations. 567pp. 6 1/2 x 9 1/4.   0-486-21076-6

BASIC WING AND AIRFOIL THEORY, Alan Pope. This self-contained treatment by a pioneer in the study of wind effects covers flow functions, airfoil construction and pressure distribution, finite and monoplane wings, and many other subjects. 1951 edition. 320pp. 5 3/8 x 8 1/2.   0-486-47188-8

SYNTHETIC FUELS, Ronald F. Probstein and R. Edwin Hicks. This unified presentation examines the methods and processes for converting coal, oil, shale, tar sands, and various forms of biomass into liquid, gaseous, and clean solid fuels. 1982 edition. 512pp. 6 1/8 x 9 1/4.   0-486-44977-7

THEORY OF ELASTIC STABILITY, Stephen P. Timoshenko and James M. Gere. Written by world-renowned authorities on mechanics, this classic ranges from theoretical explanations of 2- and 3-D stress and strain to practical applications such as torsion, bending, and thermal stress. 1961 edition. 560pp. 5 3/8 x 8 1/2.   0-486-47207-8

PRINCIPLES OF DIGITAL COMMUNICATION AND CODING, Andrew J. Viterbi and Jim K. Omura. This classic by two digital communications experts is geared toward students of communications theory and to designers of channels, links, terminals, modems, or networks used to transmit and receive digital messages. 1979 edition. 576pp. 6 1/8 x 9 1/4.   0-486-46901-8

LINEAR SYSTEM THEORY: The State Space Approach, Lotfi A. Zadeh and Charles A. Desoer. Written by two pioneers in the field, this exploration of the state space approach focuses on problems of stability and control, plus connections between this approach and classical techniques. 1963 edition. 656pp. 6 1/8 x 9 1/4.
0-486-46663-9

Browse over 9,000 books at www.doverpublications.com

Mathematics–Bestsellers

HANDBOOK OF MATHEMATICAL FUNCTIONS: with Formulas, Graphs, and Mathematical Tables, Edited by Milton Abramowitz and Irene A. Stegun. A classic resource for working with special functions, standard trig, and exponential logarithmic definitions and extensions, it features 29 sets of tables, some to as high as 20 places. 1046pp. 8 x 10 1/2. 0-486-61272-4

ABSTRACT AND CONCRETE CATEGORIES: The Joy of Cats, Jiri Adamek, Horst Herrlich, and George E. Strecker. This up-to-date introductory treatment employs category theory to explore the theory of structures. Its unique approach stresses concrete categories and presents a systematic view of factorization structures. Numerous examples. 1990 edition, updated 2004. 528pp. 6 1/8 x 9 1/4. 0-486-46934-4

MATHEMATICS: Its Content, Methods and Meaning, A. D. Aleksandrov, A. N. Kolmogorov, and M. A. Lavrent'ev. Major survey offers comprehensive, coherent discussions of analytic geometry, algebra, differential equations, calculus of variations, functions of a complex variable, prime numbers, linear and non-Euclidean geometry, topology, functional analysis, more. 1963 edition. 1120pp. 5 3/8 x 8 1/2. 0-486-40916-3

INTRODUCTION TO VECTORS AND TENSORS: Second Edition--Two Volumes Bound as One, Ray M. Bowen and C.-C. Wang. Convenient single-volume compilation of two texts offers both introduction and in-depth survey. Geared toward engineering and science students rather than mathematicians, it focuses on physics and engineering applications. 1976 edition. 560pp. 6 1/2 x 9 1/4. 0-486-46914-X

AN INTRODUCTION TO ORTHOGONAL POLYNOMIALS, Theodore S. Chihara. Concise introduction covers general elementary theory, including the representation theorem and distribution functions, continued fractions and chain sequences, the recurrence formula, special functions, and some specific systems. 1978 edition. 272pp. 5 3/8 x 8 1/2. 0-486-47929-3

ADVANCED MATHEMATICS FOR ENGINEERS AND SCIENTISTS, Paul DuChateau. This primary text and supplemental reference focuses on linear algebra, calculus, and ordinary differential equations. Additional topics include partial differential equations and approximation methods. Includes solved problems. 1992 edition. 400pp. 7 1/2 x 9 1/4. 0-486-47930-7

PARTIAL DIFFERENTIAL EQUATIONS FOR SCIENTISTS AND ENGINEERS, Stanley J. Farlow. Practical text shows how to formulate and solve partial differential equations. Coverage of diffusion-type problems, hyperbolic-type problems, elliptic-type problems, numerical and approximate methods. Solution guide available upon request. 1982 edition. 414pp. 6 1/8 x 9 1/4. 0-486-67620-X

VARIATIONAL PRINCIPLES AND FREE-BOUNDARY PROBLEMS, Avner Friedman. Advanced graduate-level text examines variational methods in partial differential equations and illustrates their applications to free-boundary problems. Features detailed statements of standard theory of elliptic and parabolic operators. 1982 edition. 720pp. 6 1/8 x 9 1/4. 0-486-47853-X

LINEAR ANALYSIS AND REPRESENTATION THEORY, Steven A. Gaal. Unified treatment covers topics from the theory of operators and operator algebras on Hilbert spaces; integration and representation theory for topological groups; and the theory of Lie algebras, Lie groups, and transform groups. 1973 edition. 704pp. 6 1/8 x 9 1/4. 0-486-47851-3

Browse over 9,000 books at www.doverpublications.com

A SURVEY OF INDUSTRIAL MATHEMATICS, Charles R. MacCluer. Students learn how to solve problems they'll encounter in their professional lives with this concise single-volume treatment. It employs MATLAB and other strategies to explore typical industrial problems. 2000 edition. 384pp. 5 3/8 x 8 1/2. 0-486-47702-9

NUMBER SYSTEMS AND THE FOUNDATIONS OF ANALYSIS, Elliott Mendelson. Geared toward undergraduate and beginning graduate students, this study explores natural numbers, integers, rational numbers, real numbers, and complex numbers. Numerous exercises and appendixes supplement the text. 1973 edition. 368pp. 5 3/8 x 8 1/2. 0-486-45792-3

A FIRST LOOK AT NUMERICAL FUNCTIONAL ANALYSIS, W. W. Sawyer. Text by renowned educator shows how problems in numerical analysis lead to concepts of functional analysis. Topics include Banach and Hilbert spaces, contraction mappings, convergence, differentiation and integration, and Euclidean space. 1978 edition. 208pp. 5 3/8 x 8 1/2. 0-486-47882-3

FRACTALS, CHAOS, POWER LAWS: Minutes from an Infinite Paradise, Manfred Schroeder. A fascinating exploration of the connections between chaos theory, physics, biology, and mathematics, this book abounds in award-winning computer graphics, optical illusions, and games that clarify memorable insights into self-similarity. 1992 edition. 448pp. 6 1/8 x 9 1/4. 0-486-47204-3

SET THEORY AND THE CONTINUUM PROBLEM, Raymond M. Smullyan and Melvin Fitting. A lucid, elegant, and complete survey of set theory, this three-part treatment explores axiomatic set theory, the consistency of the continuum hypothesis, and forcing and independence results. 1996 edition. 336pp. 6 x 9. 0-486-47484-4

DYNAMICAL SYSTEMS, Shlomo Sternberg. A pioneer in the field of dynamical systems discusses one-dimensional dynamics, differential equations, random walks, iterated function systems, symbolic dynamics, and Markov chains. Supplementary materials include PowerPoint slides and MATLAB exercises. 2010 edition. 272pp. 6 1/8 x 9 1/4. 0-486-47705-3

ORDINARY DIFFERENTIAL EQUATIONS, Morris Tenenbaum and Harry Pollard. Skillfully organized introductory text examines origin of differential equations, then defines basic terms and outlines general solution of a differential equation. Explores integrating factors; dilution and accretion problems; Laplace Transforms; Newton's Interpolation Formulas, more. 818pp. 5 3/8 x 8 1/2. 0-486-64940-7

MATROID THEORY, D. J. A. Welsh. Text by a noted expert describes standard examples and investigation results, using elementary proofs to develop basic matroid properties before advancing to a more sophisticated treatment. Includes numerous exercises. 1976 edition. 448pp. 5 3/8 x 8 1/2. 0-486-47439-9

THE CONCEPT OF A RIEMANN SURFACE, Hermann Weyl. This classic on the general history of functions combines function theory and geometry, forming the basis of the modern approach to analysis, geometry, and topology. 1955 edition. 208pp. 5 3/8 x 8 1/2. 0-486-47004-0

THE LAPLACE TRANSFORM, David Vernon Widder. This volume focuses on the Laplace and Stieltjes transforms, offering a highly theoretical treatment. Topics include fundamental formulas, the moment problem, monotonic functions, and Tauberian theorems. 1941 edition. 416pp. 5 3/8 x 8 1/2. 0-486-47755-X

Browse over 9,000 books at www.doverpublications.com

Mathematics–Logic and Problem Solving

PERPLEXING PUZZLES AND TANTALIZING TEASERS, Martin Gardner. Ninety-three riddles, mazes, illusions, tricky questions, word and picture puzzles, and other challenges offer hours of entertainment for youngsters. Filled with rib-tickling drawings. Solutions. 224pp. 5 3/8 x 8 1/2. 0-486-25637-5

MY BEST MATHEMATICAL AND LOGIC PUZZLES, Martin Gardner. The noted expert selects 70 of his favorite "short" puzzles. Includes The Returning Explorer, The Mutilated Chessboard, Scrambled Box Tops, and dozens more. Complete solutions included. 96pp. 5 3/8 x 8 1/2. 0-486-28152-3

THE LADY OR THE TIGER?: and Other Logic Puzzles, Raymond M. Smullyan. Created by a renowned puzzle master, these whimsically themed challenges involve paradoxes about probability, time, and change; metapuzzles; and self-referentiality. Nineteen chapters advance in difficulty from relatively simple to highly complex. 1982 edition. 240pp. 5 3/8 x 8 1/2. 0-486-47027-X

SATAN, CANTOR AND INFINITY: Mind-Boggling Puzzles, Raymond M. Smullyan. A renowned mathematician tells stories of knights and knaves in an entertaining look at the logical precepts behind infinity, probability, time, and change. Requires a strong background in mathematics. Complete solutions. 288pp. 5 3/8 x 8 1/2.

0-486-47036-9

THE RED BOOK OF MATHEMATICAL PROBLEMS, Kenneth S. Williams and Kenneth Hardy. Handy compilation of 100 practice problems, hints and solutions indispensable for students preparing for the William Lowell Putnam and other mathematical competitions. Preface to the First Edition. Sources. 1988 edition. 192pp. 5 3/8 x 8 1/2. 0-486-69415-1

KING ARTHUR IN SEARCH OF HIS DOG AND OTHER CURIOUS PUZZLES, Raymond M. Smullyan. This fanciful, original collection for readers of all ages features arithmetic puzzles, logic problems related to crime detection, and logic and arithmetic puzzles involving King Arthur and his Dogs of the Round Table. 160pp. 5 3/8 x 8 1/2. 0-486-47435-6

UNDECIDABLE THEORIES: Studies in Logic and the Foundation of Mathematics, Alfred Tarski in collaboration with Andrzej Mostowski and Raphael M. Robinson. This well-known book by the famed logician consists of three treatises: "A General Method in Proofs of Undecidability," "Undecidability and Essential Undecidability in Mathematics," and "Undecidability of the Elementary Theory of Groups." 1953 edition. 112pp. 5 3/8 x 8 1/2. 0-486-47703-7

LOGIC FOR MATHEMATICIANS, J. Barkley Rosser. Examination of essential topics and theorems assumes no background in logic. "Undoubtedly a major addition to the literature of mathematical logic." – *Bulletin of the American Mathematical Society.* 1978 edition. 592pp. 6 1/8 x 9 1/4. 0-486-46898-4

INTRODUCTION TO PROOF IN ABSTRACT MATHEMATICS, Andrew Wohlgemuth. This undergraduate text teaches students what constitutes an acceptable proof, and it develops their ability to do proofs of routine problems as well as those requiring creative insights. 1990 edition. 384pp. 6 1/2 x 9 1/4. 0-486-47854-8

FIRST COURSE IN MATHEMATICAL LOGIC, Patrick Suppes and Shirley Hill. Rigorous introduction is simple enough in presentation and context for wide range of students. Symbolizing sentences; logical inference; truth and validity; truth tables; terms, predicates, universal quantifiers; universal specification and laws of identity; more. 288pp. 5 3/8 x 8 1/2. 0-486-42259-3

Browse over 9,000 books at www.doverpublications.com

Mathematics–Algebra and Calculus

VECTOR CALCULUS, Peter Baxandall and Hans Liebeck. This introductory text offers a rigorous, comprehensive treatment. Classical theorems of vector calculus are amply illustrated with figures, worked examples, physical applications, and exercises with hints and answers. 1986 edition. 560pp. 5 3/8 x 8 1/2. 0-486-46620-5

ADVANCED CALCULUS: An Introduction to Classical Analysis, Louis Brand. A course in analysis that focuses on the functions of a real variable, this text introduces the basic concepts in their simplest setting and illustrates its teachings with numerous examples, theorems, and proofs. 1955 edition. 592pp. 5 3/8 x 8 1/2. 0-486-44548-8

ADVANCED CALCULUS, Avner Friedman. Intended for students who have already completed a one-year course in elementary calculus, this two-part treatment advances from functions of one variable to those of several variables. Solutions. 1971 edition. 432pp. 5 3/8 x 8 1/2. 0-486-45795-8

METHODS OF MATHEMATICS APPLIED TO CALCULUS, PROBABILITY, AND STATISTICS, Richard W. Hamming. This 4-part treatment begins with algebra and analytic geometry and proceeds to an exploration of the calculus of algebraic functions and transcendental functions and applications. 1985 edition. Includes 310 figures and 18 tables. 880pp. 6 1/2 x 9 1/4. 0-486-43945-3

BASIC ALGEBRA I: Second Edition, Nathan Jacobson. A classic text and standard reference for a generation, this volume covers all undergraduate algebra topics, including groups, rings, modules, Galois theory, polynomials, linear algebra, and associative algebra. 1985 edition. 528pp. 6 1/8 x 9 1/4. 0-486-47189-6

BASIC ALGEBRA II: Second Edition, Nathan Jacobson. This classic text and standard reference comprises all subjects of a first-year graduate-level course, including in-depth coverage of groups and polynomials and extensive use of categories and functors. 1989 edition. 704pp. 6 1/8 x 9 1/4. 0-486-47187-X

CALCULUS: An Intuitive and Physical Approach (Second Edition), Morris Kline. Application-oriented introduction relates the subject as closely as possible to science with explorations of the derivative; differentiation and integration of the powers of x; theorems on differentiation, antidifferentiation; the chain rule; trigonometric functions; more. Examples. 1967 edition. 960pp. 6 1/2 x 9 1/4. 0-486-40453-6

ABSTRACT ALGEBRA AND SOLUTION BY RADICALS, John E. Maxfield and Margaret W. Maxfield. Accessible advanced undergraduate-level text starts with groups, rings, fields, and polynomials and advances to Galois theory, radicals and roots of unity, and solution by radicals. Numerous examples, illustrations, exercises, appendixes. 1971 edition. 224pp. 6 1/8 x 9 1/4. 0-486-47723-1

AN INTRODUCTION TO THE THEORY OF LINEAR SPACES, Georgi E. Shilov. Translated by Richard A. Silverman. Introductory treatment offers a clear exposition of algebra, geometry, and analysis as parts of an integrated whole rather than separate subjects. Numerous examples illustrate many different fields, and problems include hints or answers. 1961 edition. 320pp. 5 3/8 x 8 1/2. 0-486-63070-6

LINEAR ALGEBRA, Georgi E. Shilov. Covers determinants, linear spaces, systems of linear equations, linear functions of a vector argument, coordinate transformations, the canonical form of the matrix of a linear operator, bilinear and quadratic forms, and more. 387pp. 5 3/8 x 8 1/2. 0-486-63518-X

Browse over 9,000 books at www.doverpublications.com

Mathematics–Probability and Statistics

BASIC PROBABILITY THEORY, Robert B. Ash. This text emphasizes the probabilistic way of thinking, rather than measure-theoretic concepts. Geared toward advanced undergraduates and graduate students, it features solutions to some of the problems. 1970 edition. 352pp. 5 3/8 x 8 1/2. 0-486-46628-0

PRINCIPLES OF STATISTICS, M. G. Bulmer. Concise description of classical statistics, from basic dice probabilities to modern regression analysis. Equal stress on theory and applications. Moderate difficulty; only basic calculus required. Includes problems with answers. 252pp. 5 5/8 x 8 1/4. 0-486-63760-3

OUTLINE OF BASIC STATISTICS: Dictionary and Formulas, John E. Freund and Frank J. Williams. Handy guide includes a 70-page outline of essential statistical formulas covering grouped and ungrouped data, finite populations, probability, and more, plus over 1,000 clear, concise definitions of statistical terms. 1966 edition. 208pp. 5 3/8 x 8 1/2. 0-486-47769-X

GOOD THINKING: The Foundations of Probability and Its Applications, Irving J. Good. This in-depth treatment of probability theory by a famous British statistician explores Keynesian principles and surveys such topics as Bayesian rationality, corroboration, hypothesis testing, and mathematical tools for induction and simplicity. 1983 edition. 352pp. 5 3/8 x 8 1/2. 0-486-47438-0

INTRODUCTION TO PROBABILITY THEORY WITH CONTEMPORARY APPLICATIONS, Lester L. Helms. Extensive discussions and clear examples, written in plain language, expose students to the rules and methods of probability. Exercises foster problem-solving skills, and all problems feature step-by-step solutions. 1997 edition. 368pp. 6 1/2 x 9 1/4. 0-486-47418-6

CHANCE, LUCK, AND STATISTICS, Horace C. Levinson. In simple, non-technical language, this volume explores the fundamentals governing chance and applies them to sports, government, and business. "Clear and lively ... remarkably accurate." – *Scientific Monthly.* 384pp. 5 3/8 x 8 1/2. 0-486-41997-5

FIFTY CHALLENGING PROBLEMS IN PROBABILITY WITH SOLUTIONS, Frederick Mosteller. Remarkable puzzlers, graded in difficulty, illustrate elementary and advanced aspects of probability. These problems were selected for originality, general interest, or because they demonstrate valuable techniques. Also includes detailed solutions. 88pp. 5 3/8 x 8 1/2. 0-486-65355-2

EXPERIMENTAL STATISTICS, Mary Gibbons Natrella. A handbook for those seeking engineering information and quantitative data for designing, developing, constructing, and testing equipment. Covers the planning of experiments, the analyzing of extreme-value data; and more. 1966 edition. Index. Includes 52 figures and 76 tables. 560pp. 8 3/8 x 11. 0-486-43937-2

STOCHASTIC MODELING: Analysis and Simulation, Barry L. Nelson. Coherent introduction to techniques also offers a guide to the mathematical, numerical, and simulation tools of systems analysis. Includes formulation of models, analysis, and interpretation of results. 1995 edition. 336pp. 6 1/8 x 9 1/4. 0-486-47770-3

INTRODUCTION TO BIOSTATISTICS: Second Edition, Robert R. Sokal and F. James Rohlf. Suitable for undergraduates with a minimal background in mathematics, this introduction ranges from descriptive statistics to fundamental distributions and the testing of hypotheses. Includes numerous worked-out problems and examples. 1987 edition. 384pp. 6 1/8 x 9 1/4. 0-486-46961-1

Browse over 9,000 books at www.doverpublications.com

Mathematics–Geometry and Topology

PROBLEMS AND SOLUTIONS IN EUCLIDEAN GEOMETRY, M. N. Aref and William Wernick. Based on classical principles, this book is intended for a second course in Euclidean geometry and can be used as a refresher. More than 200 problems include hints and solutions. 1968 edition. 272pp. 5 3/8 x 8 1/2.   0-486-47720-7

TOPOLOGY OF 3-MANIFOLDS AND RELATED TOPICS, Edited by M. K. Fort, Jr. With a New Introduction by Daniel Silver. Summaries and full reports from a 1961 conference discuss decompositions and subsets of 3-space; n-manifolds; knot theory; the Poincaré conjecture; and periodic maps and isotopies. Familiarity with algebraic topology required. 1962 edition. 272pp. 6 1/8 x 9 1/4.   0-486-47753-3

POINT SET TOPOLOGY, Steven A. Gaal. Suitable for a complete course in topology, this text also functions as a self-contained treatment for independent study. Additional enrichment materials make it equally valuable as a reference. 1964 edition. 336pp. 5 3/8 x 8 1/2.   0-486-47222-1

INVITATION TO GEOMETRY, Z. A. Melzak. Intended for students of many different backgrounds with only a modest knowledge of mathematics, this text features self-contained chapters that can be adapted to several types of geometry courses. 1983 edition. 240pp. 5 3/8 x 8 1/2.   0-486-46626-4

TOPOLOGY AND GEOMETRY FOR PHYSICISTS, Charles Nash and Siddhartha Sen. Written by physicists for physics students, this text assumes no detailed background in topology or geometry. Topics include differential forms, homotopy, homology, cohomology, fiber bundles, connection and covariant derivatives, and Morse theory. 1983 edition. 320pp. 5 3/8 x 8 1/2.   0-486-47852-1

BEYOND GEOMETRY: Classic Papers from Riemann to Einstein, Edited with an Introduction and Notes by Peter Pesic. This is the only English-language collection of these 8 accessible essays. They trace seminal ideas about the foundations of geometry that led to Einstein's general theory of relativity. 224pp. 6 1/8 x 9 1/4. 0-486-45350-2

GEOMETRY FROM EUCLID TO KNOTS, Saul Stahl. This text provides a historical perspective on plane geometry and covers non-neutral Euclidean geometry, circles and regular polygons, projective geometry, symmetries, inversions, informal topology, and more. Includes 1,000 practice problems. Solutions available. 2003 edition. 480pp. 6 1/8 x 9 1/4.   0-486-47459-3

TOPOLOGICAL VECTOR SPACES, DISTRIBUTIONS AND KERNELS, François Trèves. Extending beyond the boundaries of Hilbert and Banach space theory, this text focuses on key aspects of functional analysis, particularly in regard to solving partial differential equations. 1967 edition. 592pp. 5 3/8 x 8 1/2.
0-486-45352-9

INTRODUCTION TO PROJECTIVE GEOMETRY, C. R. Wylie, Jr. This introductory volume offers strong reinforcement for its teachings, with detailed examples and numerous theorems, proofs, and exercises, plus complete answers to all odd-numbered end-of-chapter problems. 1970 edition. 576pp. 6 1/8 x 9 1/4.   0-486-46895-X

FOUNDATIONS OF GEOMETRY, C. R. Wylie, Jr. Geared toward students preparing to teach high school mathematics, this text explores the principles of Euclidean and non-Euclidean geometry and covers both generalities and specifics of the axiomatic method. 1964 edition. 352pp. 6 x 9.   0-486-47214-0

Mathematics-History

THE WORKS OF ARCHIMEDES, Archimedes. Translated by Sir Thomas Heath. Complete works of ancient geometer feature such topics as the famous problems of the ratio of the areas of a cylinder and an inscribed sphere; the properties of conoids, spheroids, and spirals; more. 326pp. 5 3/8 x 8 1/2. 0-486-42084-1

THE HISTORICAL ROOTS OF ELEMENTARY MATHEMATICS, Lucas N. H. Bunt, Phillip S. Jones, and Jack D. Bedient. Exciting, hands-on approach to understanding fundamental underpinnings of modern arithmetic, algebra, geometry and number systems examines their origins in early Egyptian, Babylonian, and Greek sources. 336pp. 5 3/8 x 8 1/2. 0-486-25563-8

THE THIRTEEN BOOKS OF EUCLID'S ELEMENTS, Euclid. Contains complete English text of all 13 books of the Elements plus critical apparatus analyzing each definition, postulate, and proposition in great detail. Covers textual and linguistic matters; mathematical analyses of Euclid's ideas; classical, medieval, Renaissance and modern commentators; refutations, supports, extrapolations, reinterpretations and historical notes. 995 figures. Total of 1,425pp. All books 5 3/8 x 8 1/2.

Vol. I: 443pp.   0-486-60088-2
Vol. II: 464pp. 0-486-60089-0
Vol. III: 546pp. 0-486-60090-4

A HISTORY OF GREEK MATHEMATICS, Sir Thomas Heath. This authoritative two-volume set that covers the essentials of mathematics and features every landmark innovation and every important figure, including Euclid, Apollonius, and others. 5 3/8 x 8 1/2.   Vol. I: 461pp. 0-486-24073-8
Vol. II: 597pp. 0-486-24074-6

A MANUAL OF GREEK MATHEMATICS, Sir Thomas L. Heath. This concise but thorough history encompasses the enduring contributions of the ancient Greek mathematicians whose works form the basis of most modern mathematics. Discusses Pythagorean arithmetic, Plato, Euclid, more. 1931 edition. 576pp. 5 3/8 x 8 1/2.
0-486-43231-9

CHINESE MATHEMATICS IN THE THIRTEENTH CENTURY, Ulrich Libbrecht. An exploration of the 13th-century mathematician Ch'in, this fascinating book combines what is known of the mathematician's life with a history of his only extant work, the Shu-shu chiu-chang. 1973 edition. 592pp. 5 3/8 x 8 1/2.
0-486-44619-0

PHILOSOPHY OF MATHEMATICS AND DEDUCTIVE STRUCTURE IN EUCLID'S ELEMENTS, Ian Mueller. This text provides an understanding of the classical Greek conception of mathematics as expressed in Euclid's Elements. It focuses on philosophical, foundational, and logical questions and features helpful appendixes. 400pp. 6 1/2 x 9 1/4. 0-486-45300-6

BEYOND GEOMETRY: Classic Papers from Riemann to Einstein, Edited with an Introduction and Notes by Peter Pesic. This is the only English-language collection of these 8 accessible essays. They trace seminal ideas about the foundations of geometry that led to Einstein's general theory of relativity. 224pp. 6 1/8 x 9 1/4. 0-486-45350-2

HISTORY OF MATHEMATICS, David E. Smith. Two-volume history – from Egyptian papyri and medieval maps to modern graphs and diagrams. Non-technical chronological survey with thousands of biographical notes, critical evaluations, and contemporary opinions on over 1,100 mathematicians. 5 3/8 x 8 1/2.

Vol. I: 618pp. 0-486-20429-4
Vol. II: 736pp. 0-486-20430-8

Physics

THEORETICAL NUCLEAR PHYSICS, John M. Blatt and Victor F. Weisskopf. An uncommonly clear and cogent investigation and correlation of key aspects of theoretical nuclear physics by leading experts: the nucleus, nuclear forces, nuclear spectroscopy, two-, three- and four-body problems, nuclear reactions, beta-decay and nuclear shell structure. 896pp. 5 3/8 x 8 1/2. 0-486-66827-4

QUANTUM THEORY, David Bohm. This advanced undergraduate-level text presents the quantum theory in terms of qualitative and imaginative concepts, followed by specific applications worked out in mathematical detail. 655pp. 5 3/8 x 8 1/2.
0-486-65969-0

ATOMIC PHYSICS AND HUMAN KNOWLEDGE, Niels Bohr. Articles and speeches by the Nobel Prize–winning physicist, dating from 1934 to 1958, offer philosophical explorations of the relevance of atomic physics to many areas of human endeavor. 1961 edition. 112pp. 5 3/8 x 8 1/2. 0-486-47928-5

COSMOLOGY, Hermann Bondi. A co-developer of the steady-state theory explores his conception of the expanding universe. This historic book was among the first to present cosmology as a separate branch of physics. 1961 edition. 192pp. 5 3/8 x 8 1/2.
0-486-47483-6

LECTURES ON QUANTUM MECHANICS, Paul A. M. Dirac. Four concise, brilliant lectures on mathematical methods in quantum mechanics from Nobel Prize-winning quantum pioneer build on idea of visualizing quantum theory through the use of classical mechanics. 96pp. 5 3/8 x 8 1/2. 0-486-41713-1

THE PRINCIPLE OF RELATIVITY, Albert Einstein and Frances A. Davis. Eleven papers that forged the general and special theories of relativity include seven papers by Einstein, two by Lorentz, and one each by Minkowski and Weyl. 1923 edition. 240pp. 5 3/8 x 8 1/2. 0-486-60081-5

PHYSICS OF WAVES, William C. Elmore and Mark A. Heald. Ideal as a classroom text or for individual study, this unique one-volume overview of classical wave theory covers wave phenomena of acoustics, optics, electromagnetic radiations, and more. 477pp. 5 3/8 x 8 1/2. 0-486-64926-1

THERMODYNAMICS, Enrico Fermi. In this classic of modern science, the Nobel Laureate presents a clear treatment of systems, the First and Second Laws of Thermodynamics, entropy, thermodynamic potentials, and much more. Calculus required. 160pp. 5 3/8 x 8 1/2. 0-486-60361-X

QUANTUM THEORY OF MANY-PARTICLE SYSTEMS, Alexander L. Fetter and John Dirk Walecka. Self-contained treatment of nonrelativistic many-particle systems discusses both formalism and applications in terms of ground-state (zero-temperature) formalism, finite-temperature formalism, canonical transformations, and applications to physical systems. 1971 edition. 640pp. 5 3/8 x 8 1/2. 0-486-42827-3

QUANTUM MECHANICS AND PATH INTEGRALS: Emended Edition, Richard P. Feynman and Albert R. Hibbs. Emended by Daniel F. Styer. The Nobel Prize–winning physicist presents unique insights into his theory and its applications. Feynman starts with fundamentals and advances to the perturbation method, quantum electrodynamics, and statistical mechanics. 1965 edition, emended in 2005. 384pp. 6 1/8 x 9 1/4. 0-486-47722-3

Browse over 9,000 books at www.doverpublications.com

Physics

INTRODUCTION TO MODERN OPTICS, Grant R. Fowles. A complete basic undergraduate course in modern optics for students in physics, technology, and engineering. The first half deals with classical physical optics; the second, quantum nature of light. Solutions. 336pp. 5 3/8 x 8 1/2. 0-486-65957-7

THE QUANTUM THEORY OF RADIATION: Third Edition, W. Heitler. The first comprehensive treatment of quantum physics in any language, this classic introduction to basic theory remains highly recommended and widely used, both as a text and as a reference. 1954 edition. 464pp. 5 3/8 x 8 1/2. 0-486-64558-4

QUANTUM FIELD THEORY, Claude Itzykson and Jean-Bernard Zuber. This comprehensive text begins with the standard quantization of electrodynamics and perturbative renormalization, advancing to functional methods, relativistic bound states, broken symmetries, nonabelian gauge fields, and asymptotic behavior. 1980 edition. 752pp. 6 1/2 x 9 1/4. 0-486-44568-2

FOUNDATIONS OF POTENTIAL THERY, Oliver D. Kellogg. Introduction to fundamentals of potential functions covers the force of gravity, fields of force, potentials, harmonic functions, electric images and Green's function, sequences of harmonic functions, fundamental existence theorems, and much more. 400pp. 5 3/8 x 8 1/2.
0-486-60144-7

FUNDAMENTALS OF MATHEMATICAL PHYSICS, Edgar A. Kraut. Indispensable for students of modern physics, this text provides the necessary background in mathematics to study the concepts of electromagnetic theory and quantum mechanics. 1967 edition. 480pp. 6 1/2 x 9 1/4. 0-486-45809-1

GEOMETRY AND LIGHT: The Science of Invisibility, Ulf Leonhardt and Thomas Philbin. Suitable for advanced undergraduate and graduate students of engineering, physics, and mathematics and scientific researchers of all types, this is the first authoritative text on invisibility and the science behind it. More than 100 full-color illustrations, plus exercises with solutions. 2010 edition. 288pp. 7 x 9 1/4. 0-486-47693-6

QUANTUM MECHANICS: New Approaches to Selected Topics, Harry J. Lipkin. Acclaimed as "excellent" (*Nature*) and "very original and refreshing" (*Physics Today*), these studies examine the Mössbauer effect, many-body quantum mechanics, scattering theory, Feynman diagrams, and relativistic quantum mechanics. 1973 edition. 480pp. 5 3/8 x 8 1/2. 0-486-45893-8

THEORY OF HEAT, James Clerk Maxwell. This classic sets forth the fundamentals of thermodynamics and kinetic theory simply enough to be understood by beginners, yet with enough subtlety to appeal to more advanced readers, too. 352pp. 5 3/8 x 8 1/2. 0-486-41735-2

QUANTUM MECHANICS, Albert Messiah. Subjects include formalism and its interpretation, analysis of simple systems, symmetries and invariance, methods of approximation, elements of relativistic quantum mechanics, much more. "Strongly recommended." – *American Journal of Physics.* 1152pp. 5 3/8 x 8 1/2. 0-486-40924-4

RELATIVISTIC QUANTUM FIELDS, Charles Nash. This graduate-level text contains techniques for performing calculations in quantum field theory. It focuses chiefly on the dimensional method and the renormalization group methods. Additional topics include functional integration and differentiation. 1978 edition. 240pp. 5 3/8 x 8 1/2.
0-486-47752-5

Browse over 9,000 books at www.doverpublications.com

Physics

MATHEMATICAL TOOLS FOR PHYSICS, James Nearing. Encouraging students' development of intuition, this original work begins with a review of basic mathematics and advances to infinite series, complex algebra, differential equations, Fourier series, and more. 2010 edition. 496pp. 6 1/8 x 9 1/4. 0-486-48212-X

TREATISE ON THERMODYNAMICS, Max Planck. Great classic, still one of the best introductions to thermodynamics. Fundamentals, first and second principles of thermodynamics, applications to special states of equilibrium, more. Numerous worked examples. 1917 edition. 297pp. 5 3/8 x 8. 0-486-66371-X

AN INTRODUCTION TO RELATIVISTIC QUANTUM FIELD THEORY, Silvan S. Schweber. Complete, systematic, and self-contained, this text introduces modern quantum field theory. "Combines thorough knowledge with a high degree of didactic ability and a delightful style." – *Mathematical Reviews.* 1961 edition. 928pp. 5 3/8 x 8 1/2. 0-486-44228-4

THE ELECTROMAGNETIC FIELD, Albert Shadowitz. Comprehensive undergraduate text covers basics of electric and magnetic fields, building up to electromagnetic theory. Related topics include relativity theory. Over 900 problems, some with solutions. 1975 edition. 768pp. 5 5/8 x 8 1/4. 0-486-65660-8

THE PRINCIPLES OF STATISTICAL MECHANICS, Richard C. Tolman. Definitive treatise offers a concise exposition of classical statistical mechanics and a thorough elucidation of quantum statistical mechanics, plus applications of statistical mechanics to thermodynamic behavior. 1930 edition. 704pp. 5 5/8 x 8 1/4.
0-486-63896-0

INTRODUCTION TO THE PHYSICS OF FLUIDS AND SOLIDS, James S. Trefil. This interesting, informative survey by a well-known science author ranges from classical physics and geophysical topics, from the rings of Saturn and the rotation of the galaxy to underground nuclear tests. 1975 edition. 320pp. 5 3/8 x 8 1/2.
0-486-47437-2

STATISTICAL PHYSICS, Gregory H. Wannier. Classic text combines thermodynamics, statistical mechanics, and kinetic theory in one unified presentation. Topics include equilibrium statistics of special systems, kinetic theory, transport coefficients, and fluctuations. Problems with solutions. 1966 edition. 532pp. 5 3/8 x 8 1/2.
0-486-65401-X

SPACE, TIME, MATTER, Hermann Weyl. Excellent introduction probes deeply into Euclidean space, Riemann's space, Einstein's general relativity, gravitational waves and energy, and laws of conservation. "A classic of physics." – *British Journal for Philosophy and Science.* 330pp. 5 3/8 x 8 1/2. 0-486-60267-2

RANDOM VIBRATIONS: Theory and Practice, Paul H. Wirsching, Thomas L. Paez and Keith Ortiz. Comprehensive text and reference covers topics in probability, statistics, and random processes, plus methods for analyzing and controlling random vibrations. Suitable for graduate students and mechanical, structural, and aerospace engineers. 1995 edition. 464pp. 5 3/8 x 8 1/2. 0-486-45015-5

PHYSICS OF SHOCK WAVES AND HIGH-TEMPERATURE HYDRO DYNAMIC PHENOMENA, Ya B. Zel'dovich and Yu P. Raizer. Physical, chemical processes in gases at high temperatures are focus of outstanding text, which combines material from gas dynamics, shock-wave theory, thermodynamics and statistical physics, other fields. 284 illustrations. 1966–1967 edition. 944pp. 6 1/8 x 9 1/4.
0-486-42002-7